THE SCHUPMANN TELESCOPE

The Story, Design, Construction and Use of a Neglected Telescope Type

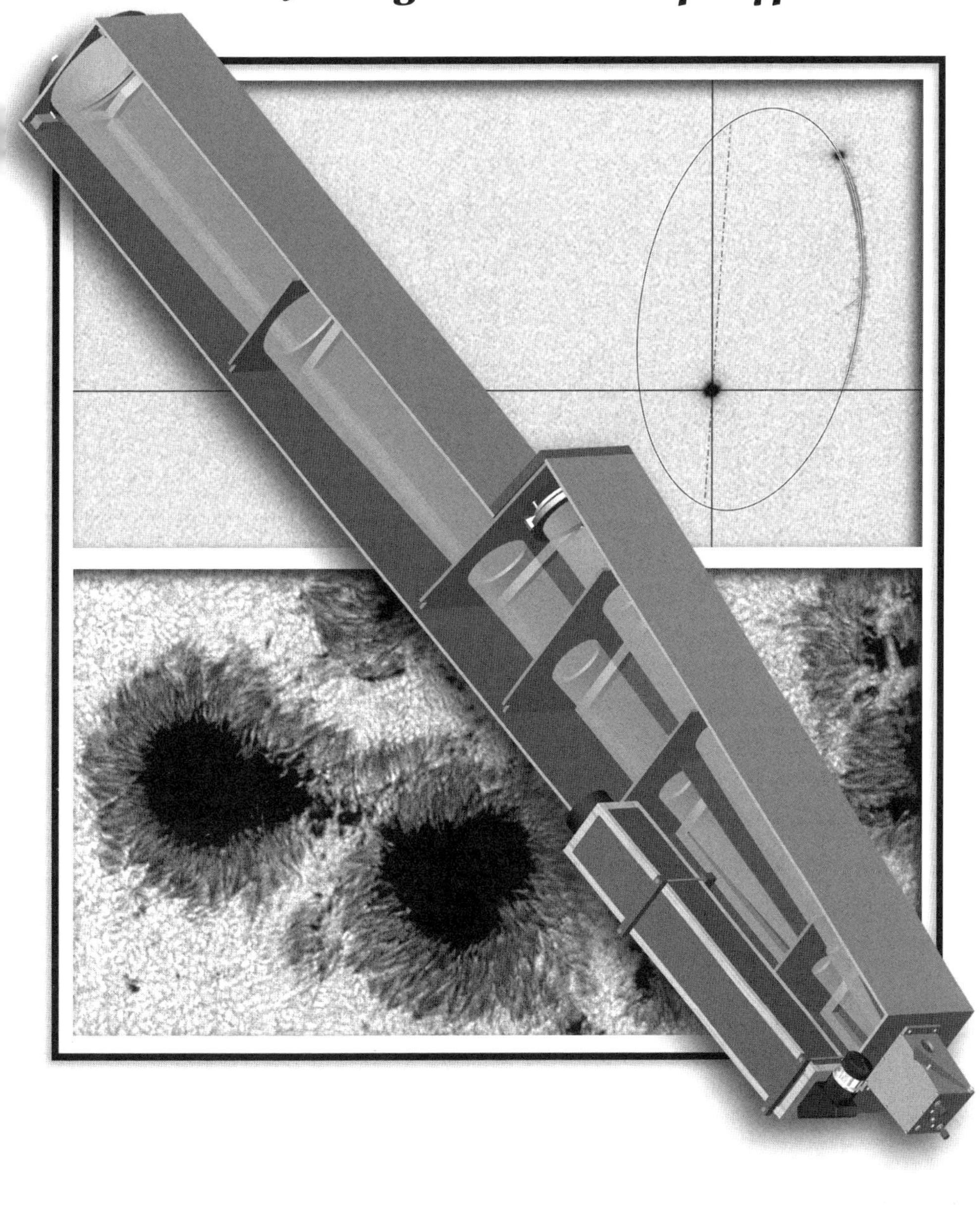

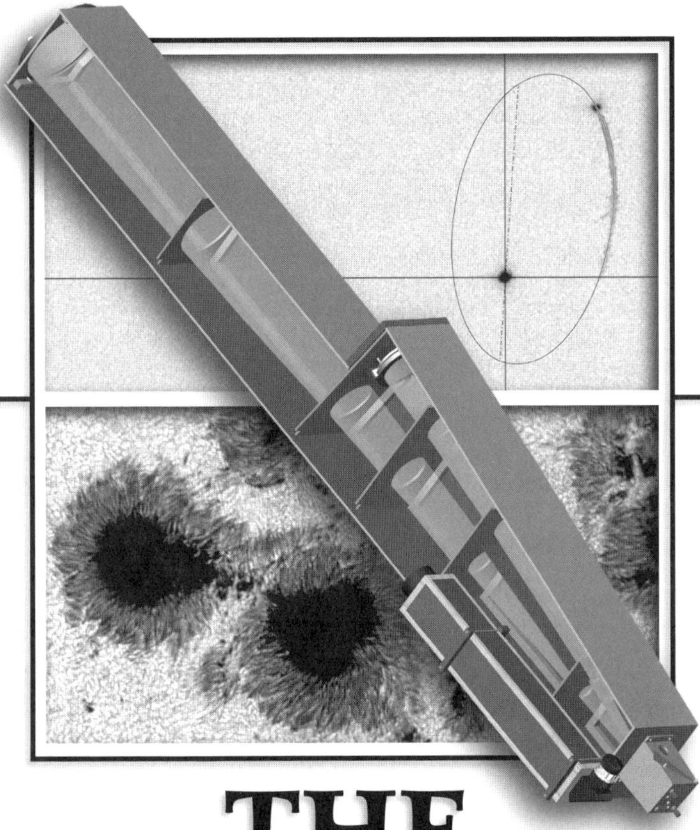

THE SCHUPMANN TELESCOPE

The Story, Design, Construction and Use of a Neglected Telescope Type

JAMES DALEY

Published by

Willmann-Bell, Inc.

P.O. Box 35025 • Richmond, VA 23235 • Toll Free 1-800-825-7827 • 804-320-7016 • Fax 804-272-5920

www.willbell.com

Published by Willmann-Bell, Inc., P.O. Box 35025, Richmond, Virginia 23235

Printed in the United States of America

Library of Congress Cataloging in Publication Data

Daley, James, 1936-

The Schupmann telescope / James Daley.

p. cm.

Includes bibliographical references and index.

ISBN-13: 978-0-943396-59-0

1. Schupmann, Ludwig, 1851–1920. 2. Schupmann medial telescopes. 3. Refracting telescopes--Design and construction--Amateur's manual.

QB88 .D35 2006

681/.4123 22 2006049799

07 08 09 10 11 9 8 7 6 5 4 3 2 1

Table of Contents

Preface

The modern amateur telescope maker is indeed lucky. The choice of top-rate building materials, glass types, instrument designs and technical resources has never been better. Inexpensive ray trace programs abound, allowing the Telescope Nut (TN) to enter the realm of optical design or simply adjust an existing recipe to test an idea. The amateur telescope maker (ATM) finds great satisfaction in the construction process and after a spell of "cosmic touring" with the completed scope, the urge grows to begin a new, usually bigger, instrument. The ATM dreams of reduced tube currents, higher wave-front correction, and perhaps a bigger field. The night sky is their "test track", their "proving ground"—a close double separated, a faint NGC object spotted. Although a few bridge the gap, contributing to astronomical science is typically not in the cards for these builders. The instrument and its perfection alone is their passion and life hobby. It is from this dedicated group, down through the years—both amateur and professional—that so much innovative design has emerged to the benefit of all astronomers. We applaud this fine tradition. It has provided the underpinnings, the very tools used to help reveal the characteristics of the universe.

How can the joy of telescope making be explained? Is it the wonderful mixture of skills—designing the instrument, machining the cells, building the tube, and grinding and polishing the beautiful optics? Or is it something more magical—the promise of an even better view than with the last instrument, or the lure of discovery? Can it be the camaraderie of like-minded friends and the joys shared? Is it the technical problems solved, the anguishing moments overcome? Just what drives the ATM? I think it's all of the above and more. We are a rare bunch of birds for sure; just ask any of our non-TN acquaintances!

Schematically, the visual telescope seems so simple: a "black box" that transforms a large beam of light into a small beam, by way of an eyepiece, just the right size in this example, to exactly fit our eye's pupil. The ratio of the beam diameters is called the magnification and the ratio of the areas is suggested in the wonderfully delightful phrase "light-gathering power" or perhaps "light-grasp". We need only look into the eyepiece and the illusion is complete: the stars appear closer to us (wider apart) and seem brighter, bringing thousands of unseen stars into view. We have, in effect, simply scaled-up the human visual detection system: the eyeball! And when the telescope is used with film or a CCD, the system is conceptually even simpler since the instrument is just a big camera.

Why are there scores of telescope designs to accomplish the same basic function? Reasons can be found in the literature, but reduced to a simple answer we can say that the observational requirements in each astronomical discipline are not exactly the same nor are personal preferences. Each category of instrument has special advantages; *i.e.,* shorter tube, greater coma-free field, better color correction, lower focal ratio, less light scatter, more portability, less cost and on it goes. In the case of disciplines for instance: solar telescopes do poorly on faint galaxies and "light bucket" Newtonians haven't a chance as a white-light coronagraph; nor are rich field telescopes at their best on the planets. A 60mm refractor would be a more than difficult choice with which to visually observe M 109—an example of how the required aperture can drive instrument design and, in fact, can have more to do with instrument choice than any other requirement. Up to about 300 mm aperture the designs to satisfy various requirements seem endless. However, as the required aperture increases, even slightly, over this value the "simple" Newtonian or Cassegrainian forms and their many variations take front and center on the ATM's workbench.

The type of telescope described in this book has been my ATM passion for over 40 years. I've never found a small telescope design so versatile, with so many possibilities and so delightful to use under the stars. In 1984 I self-published the book *Amateur Construction of Schupmann Medial Telescopes*, which led to the construction of many notable and beautifully successful Schupmann's, thus fulfilling a dream. Many years have passed and it's time to bring the subject up to date.

James A. Daley, Jr.
Springfield Telescope Makers
December 2006

Foreword

Rare is the telescope design that languishes in near total obscurity and yet displays such exquisite performance as that invented by Ludwig Schupmann. His remarkable discovery, at the end of the 19th century, was that it is possible to achieve color correction equivalent to an all-reflecting telescope with a single-lens objective followed by a smaller corrector lens-mirror combination placed beyond the focus of the objective. However, to this day very few telescopes based on his principle have been made and most of those have been by advanced amateurs. Unlike the Newtonian telescope, the Schupmann principle has shown an ability to delight optical designers with its many variations; there is an ideal solution for planetary and lunar observing, there is a perfect solution for coronagraphic optics which can reveal solar prominences without expensive narrowband filters, and there is an all-spherical design.

It continues to offer surprises—for example, only in recent years has a wide-field low-*f*/number variation been discovered that is capable of outperforming existing astrographic cameras. If Ludwig Schupmann is the inventor of this remarkable telescope, Jim Daley could be called its perfector. No one has built more models or dedicated more of his life to promoting this instrument. His production line has brought to light a 7-inch in 1969, a 4-inch in 1973, a 6-inch in 1984, a 7-inch in 1988, an 8-inch in 1990, a 9.4-inch in 2002 and as this book goes to press in 2006 a 14.1-inch! His greatest pleasure with the Schupmann telescope is using it to measure double stars and having his results consistently accepted by professional publications.

This is a backwater astronomical pursuit that attracts few followers, but an important one, because it allows astronomers to improve their understanding of the relationship between stellar masses, measurable only by observing binary stars, and spectral classification. To help promote interest in and understanding of this telescope design Jim decided to publish a bulletin which appeared as three issues between 1966 and 1971. Later, it became clear that all the information known about the Schupmann telescope, from historical to fabrication to operation, should be brought together under one cover which he titled, *Amateur Construction Of Schupmann Medial Telescopes*, privately published in 1984.

In the present book Jim weaves a thorough and detailed account of this versatile telescope design, including the most recent innovations. The latest chapter in the Schupmann story takes us back to Europe and a variation of the Schupmann principle in which the corrector lens-mirror is placed before the focus of the single objective lens. Even though Schupmann claimed in his patent that this variation was capable of excellent color correction, the earliest attempts at fabricating such a design were unsuccessful due to poor workmanship. Klaas Honders, an ATM in Amsterdam, changed all that when he discovered a design with excellent performance, yet he thought it impractical to build and put it aside for over ten years. Finally, using more advanced optical design software, he found he was able to vastly improve his earlier versions to the extent that, with only three elements, the new designs exceeded the performance of most astrographs, a remarkable discovery. One is drawn to ponder what other designs lay hidden in the concept—the exploration has barely begun.

Berton C. Willard
Historian, The Springfield Telescope Makers and
author of *Russell W. Porter, Arctic Explorer, Artist, Telescope Maker*

Acknowledgements

Many people helped to make this book possible. The names of Schupmann makers and their contributed photos and comments can be found throughout the book as well as those who suggested innovative ideas. I especially thank fellow Springfield Telescope Makers members, Bert Willard—for his untiring interest in the medial, for writing the Foreword, and for the enormous amount of ray trace work he has provided me over the years—and Scott Milligan, who played a major leadership role in bringing the club's 13-inch medial into being, thus greatly increasing the awareness of this design in the ATM community. Scott kindly contributed a practical experience story of the 13-inch for this book. Chapter 9 on the brachymedial is made complete with the generous help and contributions of: Rainer Ortmann, restorer and curator of the Rathenow brachymedial; Roger Ceragioli, for his abiding interest in the Schupmann and his many photos and ideas regarding the brachymedial; Matt Considine, for unearthing historically important articles in the *English Mechanic*; Will Weidner, my steamboating buddy, who skillfully generated the many perspective drawings seen throughout the book, including the cover illustration; Clifton Ashcraft, whose contributions in the early days shaped the modern Schupmann and whose sharp, analytical, view of the instrument continues to this day; Oliver Wachter and Klaus Jünemann, for pictures of Schupmann's street light fixtures seen in this book; and Uli Mayer, who kindly contributed his find of historic brachymedial drawings by Schupmann.

I gratefully acknowledge those who gave so much time and encouragement and have now passed away, among them: Rev. Edwin Olson, an extraordinary ATM who was a true pioneer of medial innovations and who introduced me to the design; Anton Kutter, a great champion and builder of medials and other advanced designs, who filled in so many historical gaps concerning the development of the Schupmann in Germany; and most recently, as my manuscript was reaching final form, Klaas Honders who brought the neglected brachymedial of age and who generously contributed two beautiful designs to this book.

I have been warmly encouraged throughout this project by Perry Remaklus of Willmann-Bell, Inc. This sort of narrow-interest book is really a labor of love for all concerned and I thank Perry and his colleagues, especially John D. Koester, for their genuine interest in the Schupmann and expert knowledge of the amateur telescope making field. This made my work so much easier.

Introduction

As I began to envision a book about the Schupmann telescope, many ideas of how to slant its emphasis were tossed about. Writing another telescope making tome which described all the steps in grinding and figuring lenses and mirrors, would simply repeat what is already well presented in so many other volumes. Done that way, the book would be straightforward: present a single design and, in great detail, describe its particular optical characteristics, construction, and alignment and leave it at that.

In the end I decided to follow a somewhat different path—that is, to tell a much broader story of the Schupmann instrument especially aimed at the advanced TN and serious observer who not only wishes to experience the building of a fine instrument, but may also want to employ his new medial to contribute to astronomical knowledge by sharing data with their fellow amateur and professional colleagues.

The reader will find no mathematical optical formulae here, no Bessel functions or fancy theories about atmospheric seeing—just a book about a fascinating telescope form, its historical development, various design prescriptions and their optical performance, special construction details, and something about its many possibilities. Optical theorists will, no doubt, find the book a bore. However, on closer inspection they may suddenly be struck with a neat thought, a thought that could open a bright new path for exploration. For the student of telescopes and their historical development, and in particular those historians of unusual telescope types, it is hoped that this book may find a home on their library shelves. There is surely something here to tickle their fancy.

All the ray trace work presented here was performed with OSLO-LT, software that is freely downloadable from the Internet. Although the learning curve was steep—especially for someone with no formal education in optical design like myself—with the gracious help of fellow STM member, Bert Willard, I soon managed to create and analyze all sorts of systems, tilts and all! I deeply thank the folks of OSLO for providing such a powerful design tool for the student optical enthusiast.

I especially want to hear of the plans of amateurs who decide to build a Schupmann medial or its distant cousin, the brachymedial. Errors do creep into any publication despite going over and over the manuscript. If you do build to a supplied prescription, and you have no way to check the design, please let me know and I will double check it and send you any corrections. I can be contacted through the Stellafane website.

In the end we are, in large part, simply messengers carrying the lore. Do have a wonderful experience with your Schupmann.

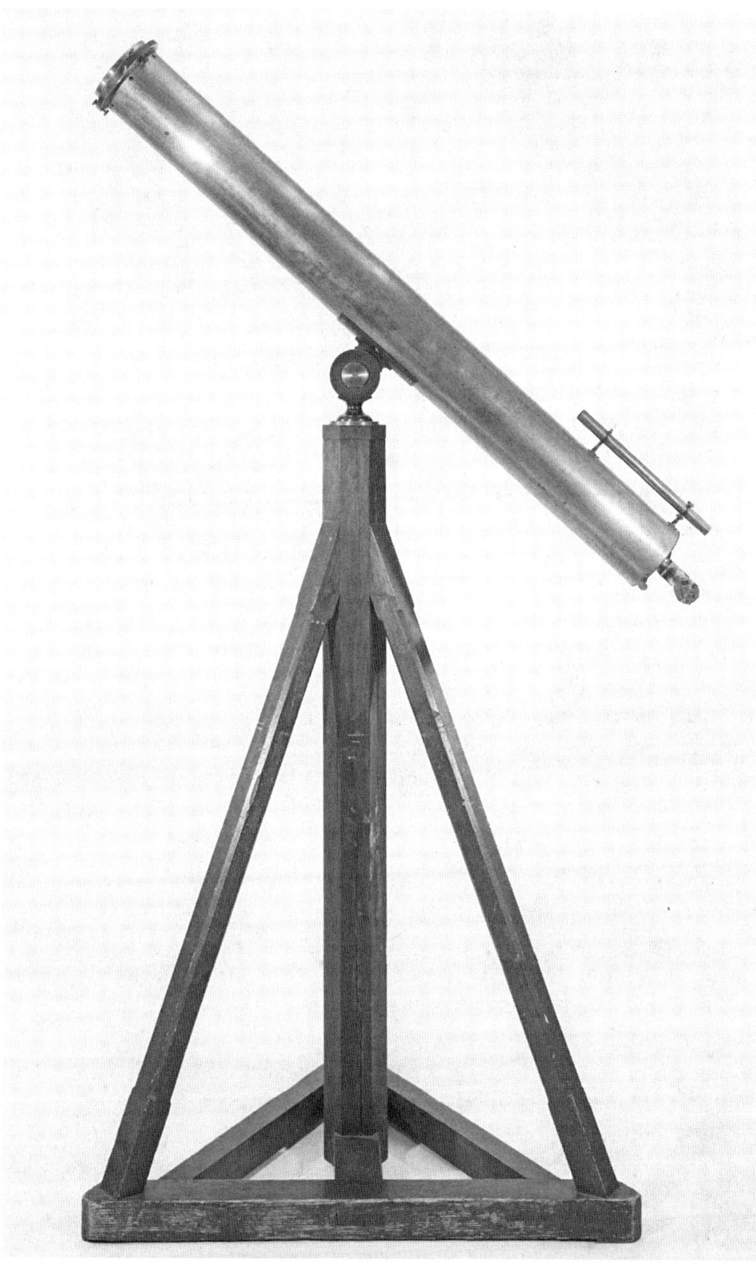

Fig. 1.1 *Photo of the worlds first medial telescope. This 5-inch f/12.8 instrument was built by Professor Ludwig Schupmann in 1897 as a pre-patent proof-of-principle model. Schupmann's tiny color corrector of only 5/9-inch diameter is mounted in his "elbow" tailpiece arrangement, consistent with the patent drawings and most future German medials. Courtesy Deutsches Museum, Munich. All Rights Reserved.*

Chapter 1
The History of the Schupmann Telescope

1.1 Ludwig Schupmann and the Early German Medials

There can hardly be a professional optical designer or amateur telescope maker who, by now, is not familiar with the principle of the Schupmann telescope. However, little is known of the inventor. This historical account begins with a fragmented story based on the author's meager records of the man himself, a little about two of his friends and the medials of his homeland.

Fig. 1.2 *Ludwig Schupmann, 1851–1920. Photo courtesy Edgar Lüüs.*

Born on January 23, 1851 in Geseke, Germany, Ludwig Ignaz Schupmann was the sixth and last child of Dr. Adolf Schupmann and Therese (Beckers) Schupmann. Ludwig's father was a medical doctor, a career followed by Ludwig's older brother Xaver. Ludwig showed, even in his early youth, a strong interest in astronomical goings-on and in particular the technical side of the instrumentation. Although his piercing analytical mind was evident to all those around him, his overly modest character probably reduced the impact of many of his future achievements. He simply was not one to shout from the highest rooftop his truly great new ideas. Kutter remarked of Schupmann: "His life reads like a romance, is a romance, but one without a happy end." Ludwig Schupmann died at home at a relatively young age of 69 of an acute bladder infection. He was given last sacraments surrounded by his nearest family members. Schupmann never married.

After earning his degree at the Technical University of Aachen (Rhineland) he worked in the field of architectural design specializing in public buildings. His achievements in this field resulted in many top awards at design competitions. His ornate designs of street lighting "Kandelaber" are still used in Germany today and may represent the first instance of cut-off lighting as the globes themselves were

Fig. 1.3 *Ludwig Schupmann's Kandelaber street lights are still in use today. The double fixture shown at the left is found at the Brandenburg Gate in Berlin.*

internally silvered on their upper hemispheres! (See Figure 1.3.) Ludwig was an accomplished artist and it is remarkable how his life, in many ways, paralleled Russell W. Porter's. Ludwig's pencil sketches, watercolors and pastel work are of the highest standard. At age 36 he was appointed to a professorship at the same university from which his degree was earned.

It is not clear when Schupmann hit upon the optical principle for which he now is famous. His struggle to free the refractor of its devastating color defects and limits on aperture, due not only to chromatic aberration, but also flexure and support problems especially associated with the flint element, finally led him to investigate the 1850 dialyt design of Georg Simon Plössl (1794–1868). This form of refractor offered the hope of correcting a large single-element objective with a doublet lens positioned about halfway down the tube. Failing a satisfactory solution, Schupmann abandoned the dialyt design. The telescope ideas[1] of Caleb Smith (1683–1775), Alphonse Mangin (1825–1885), and William Hamilton (1814) had design elements which hinted at more promising possibilities and were probably known to Schupmann. In any event, Schupmann hit upon his unique invention which offered a solution for color correction unheard of at the time for which he was granted German patent No. 108181, dated October 22, 1898. A very nice drawing in the patent shows the two now familiar arrangements, the brachymedial and the medial. Curiously nonastronomical, the patent also describes the use of his color correction principle for a microscope objective, also well shown in the patent drawing (see Figure 1.4). In 1899 he published his breakthrough

[1] Riekher, Rolf, *Fernrohre und ihre Meister*, Verlag Technik, Berlin,1990. pp. 230–236.

L. SCHUPMANN IN AACHEN.

Optische Vorrichtung zur Achromatisirung eines nicht achromatischen Objectivs.

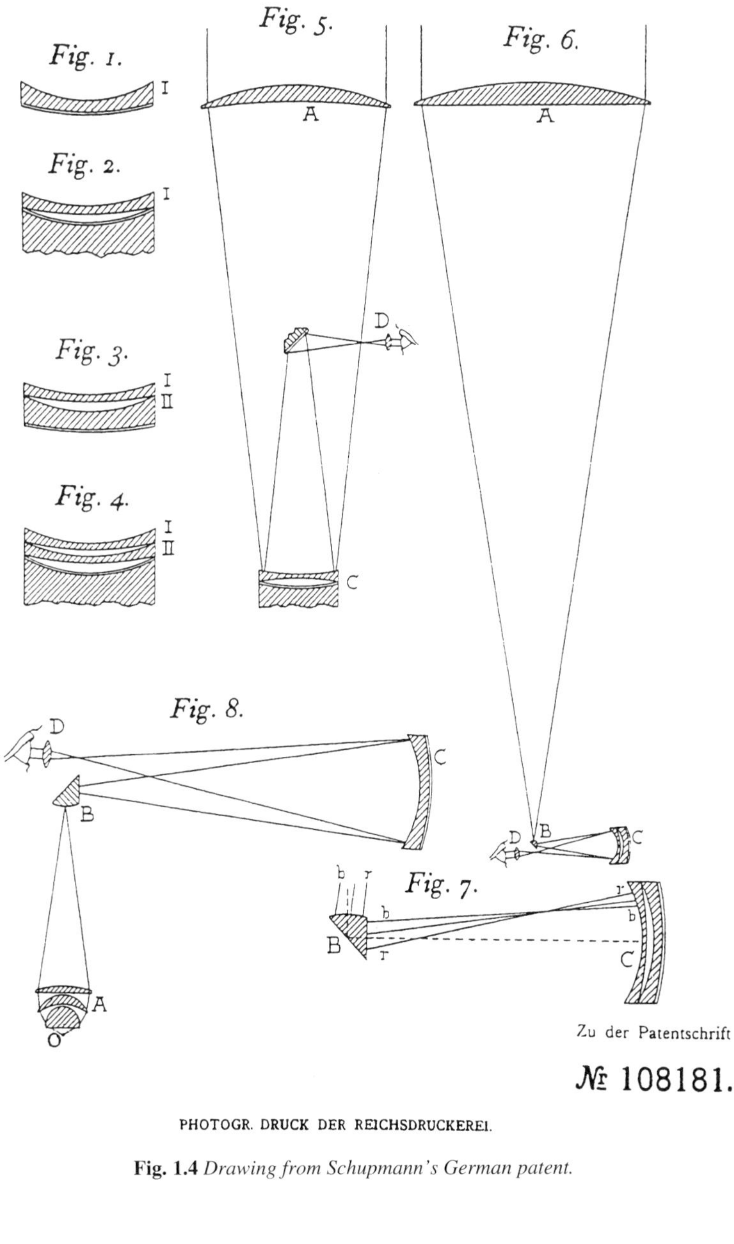

Zu der Patentschrift

№ 108181.

PHOTOGR. DRUCK DER REICHSDRUCKEREI.

Fig. 1.4 *Drawing from Schupmann's German patent.*

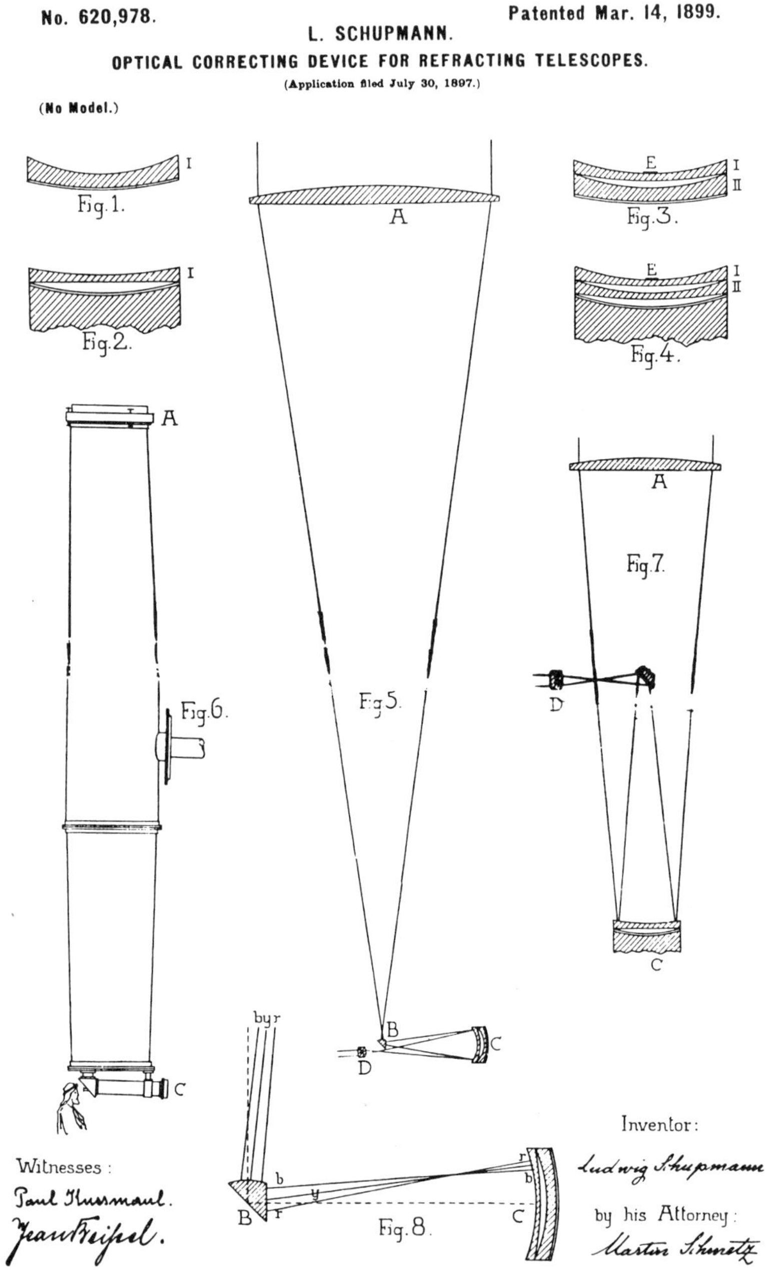

Fig. 1.5 *Drawing from Schupmann's U.S. patent.*

work titled *Die Medial-Fernrohre.*[2] This 146 page book describes various solutions for nearly complete color correction for apertures up to 60 inches. Looking at the fine line drawings in this work, one cannot help thinking that they could only have been done only by the hand of Schupmann. It was in that year that Ludwig was granted a U.S. patent for his telescope designs (see Figure 1.5). He applied for the patent on July 30, 1897 while living in Aix-la-Chapelle (Aachen), Germany. The patent title is "Optical Correcting Device For Refracting-Telescopes" No. 620,978, dated March 14,1899.

An interesting early (1900) refutation to Schupmann's medial telescope inventions appeared in *The Observatory*[3] where it was claimed that the work of two members of the Toronto Astronomical Society, the brothers Z. M. and J. R. Collins, predated Schupmann's work. Their instrument called the "Monoplane Achromatic Telescope" was an outgrowth of their dialyt experiments, and was clearly realized in 1896. It was of the brachymedial form and had a 4-foot focal length enclosed in a 2-foot tube and had a clear aperture of 4 inches. The closely related Hamiltonian design was, of course, available much earlier. The important point here is that the refutation blanketed the *medial* type which, of course, is quite a different instrument from just about every standpoint. The account tells of the submission of drawings and specifications to such optical luminaries as Lord Kelvin, Professor J.A. Brashear and Dr. H.C. Vogel. Although nothing ever came of their claim, the brothers apparently went on to construct an 8-inch version of their "Monoplane Achromatic Telescope" at the encouragement of Vogel and others. Vogel, of the Astrophysical Observatory, Potsdam, analyzed the design and reported: "a workable objective is now shown which, as the reasoning proves, unites in an excellent manner the rays from red to violet, better indeed than is the case with an objective of the usual construction."

Schupmann heard of Musson's claims on behalf of the Collins brothers, and wrote a most interesting and historically important rebuttal to Musson's contention. The letter appeared on pages 229–230 of the October 19, 1900 issue of the *English Mechanic*. The author feels that the details of the medial's development contained in Schupmann's letter are so significant and delightfully written, that it is included in its entirety here. The letter is titled "The Medial Telescope" and we read:

> (44103)-Number 1900 of the "EM.," p.61, contains a communication by Mr. W. B Musson, secretary of the Toronto Astronomical Society, in which Mr. Musson appears to claim the priority of invention of a new kind of telescope, called by me "Brachymedials" not for me, but for two members of his society, Messrs. Z. M. and J. R. Collins, of Toronto.
>
> He did not think fit to favour me with a copy of said communication, and it was by mere accident that I became acquainted with it. As the drift of Mr. Musson's communication may be misunderstood, I think it right to submit to scientists the following

[2] Schupmann, Ludwig, *Die Medial-Fernrohre*, Eine neue konstruktion für grosse astronomische instrumente, Leipzig. 1899.

[3] W. B. Musson, *The Observatory*, Correspondence, September, 1900, pp. 350–352.

facts, fearing silence on my part might be misinterpreted.

1. The various kinds possible of the new instruments and the conditions of removing the secondary spectrum in the same, were found out by me as early as the year 1892—that is, at the time when Messrs. Collins were not yet thinking of the construction of such an instrument. In 1893, that is to say, three years before Messrs. Collins, I put together the first "Brachymedial," as may be read in my book published in the beginning of 1899. The two lenses necessary for the purpose (a single plane convex objective lens and a smaller concave meniscus) I had already purchased from Herr Venneman, optician at Aix-la-Chapelle, before the end of 1892.

2. Regarding "confidential communications" to competent gentlemen, to which Mr. Musson seems to attribute great importance, in that respect also I forestalled Messrs. Collins. While the latter first communicated their ideas to Lord Kelvin, Prof. Brashear, and to Prof. Vogel in May, 1897, I had already laid my complete theory before Prof. W. Wien (the successor of Roentgen at Wuerzburg), in November, 1896—that is to say, fully six months earlier. Prof. Wien entirely approved of my theory, and encouraged me to proceed with the construction of the "medial." The two small concave meniscuses, which are a most characteristic part of the invention, were ordered from the firm of Voightländer u. Sohn, at Braunschweig, as early as April 12, 1897. The whole instrument was finished by June 1897.

3. This "medial," the most valuable type of the new instruments (in which the light reaches the concave correcting system and the concave mirror surface after having passed through a prism placed close to the focus of the object-glass) has not been hit upon by Messrs. Collins at all.

4. And last I desire to call attention to a fact, compared to with those mentioned before are of slight significance. The only publication referring to the subject has been by me. ("Die Medial Fernrohre," Leipzig, B. G. Teubner.1899). Now it is a custom universally prevalent among scientists that the priority of a discovery belongs scientifically to him who first publishes in a scientific manner the principle and details of an invention. Nothing of the kind has been attempted by Messrs. Collins. The addresses to the Astronomical Society of Toronto are anything rather than publications of the invention. The most essential part of the invention, the concave mirror surface, is not so much as hinted at in the address.

While unable to admit Messrs. Collins claim to priority I have no hesitation in acknowledging that they have later than I, but quite independently discovered the less valuable type, the "Brachymedial." In conclusion, I should like to know what improvement which Messrs. Collins claim to have made in the Brachymedial consists in.

It may interest the readers of the *English Mechanic* that since June 5, 1900, a medial "of 12in. Aperture and 15ft. focal length" has been placed in the "Urania" Observatory at Berlin. The instrument was constructed according to my calculations by the well-known firm of Reinfelder u. Hertel at Muenchen. The instrument produces entirely achromatic images even of the brightest celestial objects. Double stars of 0.6 distance, as, for example, χ Aquila, are separated by it without any difficulty.

I hope you will give me a chance of stating my case to your readers by publishing this letter in your valuable paper. Prof. L. Schupmann. Aix-la-Chapelle, Oct, 10.

Schupmann's letter seems to have put to rest any prior claims to his invention as nothing more is heard on the matter.

Getting back to Schupmann's work: the first medials constructed were two pre-patent models of 127mm aperture, *f*/12.8, instruments one a medial of the "elbow" layout, the second a brachymedial. It is clear from Schupmann's forgoing rebuttal that he completed the brachymedial in 1893 and the medial in 1897. It is also evident that he did not grind and polish the elements for these prototypes. After careful reading of the patent the author is inclined to believe that the same objective was used in both arrangements as proof of principle for the patent. From the patent we read: "An instrument with an object-glass of five inches clear opening, 64-inches focal length, and 5/9ths of an inch opening of the mirror has been built, and most satisfactory results have been obtained with same." By "mirror" he was speaking of the corrector. Later in the patent and speaking of the brachymedial we read: "I have tested the modification in Fig.7 (see Figure 1.5) which only by means of the object-glass and my correcting device unites the light perfectly in the focus." The whereabouts of these experimental models is unknown. A photograph showing this first medial was located in a Munich museum (see Figures 1.1) from which a sketch (see Figure 1.6) was made by the author to better show the corrector end. The first large medial was a 333mm-aperture version manufactured by Reinfelder and Hertel in 1900. This instrument, although moderately successful, did not typify the best of Schupmann's designs. It was later destroyed in bombing raids.

Among Ludwig's friends was a very special person, Philipp Fauth. Schupmann provided Fauth with a design for a greatly improved version of the medial and, in this connection, the reader may wish to see *Sky and Telescope* for November, 1959, pp. 21–24, "Philipp Fauth and the Moon" by Hermann Fauth. Fauth's selenographic work remains, even today, an amazing achievement. The 385mm Fauth medial, built by Merz in 1911, was of exceptional quality; and a smaller 323mm copy, also by Merz in 1921, underwent extensive testing and was pronounced perfect by astronomers[4] of the Konigstuhl-Observatory (Heidelberg) and is still in use in Stuttgart.[5] Anton Kutter, who regarded Schupmann as his mentor, is known throughout the ATM community for his "Schiefspiegler" (Neo-Brachyt), a brilliant expansion on the 1876 "Brachytelescope" design of Forster and Fritsch. Kutter observed with the Fauth medial in the years 1932 until Fauth's death in 1941.[6] Schupmann died in the town of his birth on October 2, 1920 and, thus lived to see the realization of this specific design (Fauth Instrument). This design, modified by Schupmann for single element compensation, was finally published 10 months after his death without the benefit of proofreading and was filled with confusing typographical errors.[7] This, combined with Schupmann's own

---

[4] Mündler, M., "Über ein 12-zölliges Medial," *Astronomishe Nachrichten*, No. 5223, Vol. 218, April 1923, pp. 235–240.

[5] Wolter, Cristian and Merz, Rainer, "Gleanings For ATM's" conducted by Roger W. Sinnott, "The Neglected Schupmann Refractor", *Sky and Telescope*, March, 1983, pp. 273–278.

[6] This instrument was ravaged in the after-war difficulties.

[7] Schupmann, Ludwig,"Zeitschrift für Instrumentarkude" XLI Jahrgang, pp. 212–219.

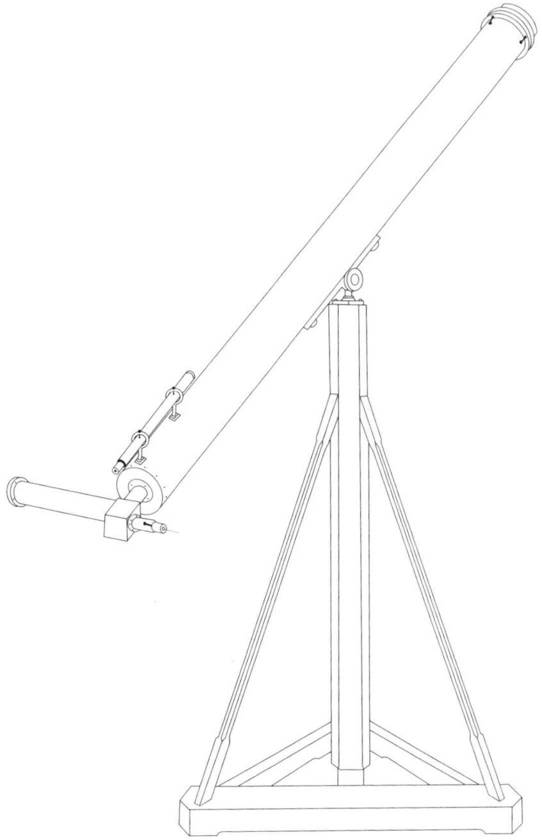

Fig. 1.6

mathematical symbols, did little to advance the medial form.

It is from the author's private communication with the late Anton Kutter that much of the information given here was gleaned. His enthusiasm for the renewed interest in the Schupmann telescope in the United States was very evident in his letter. Kutter constructed three medials, of which the first two were exact models of the Fauth instrument. The first, the "Berta Kutter Medial", of 122mm aperture was finished in 1936. The second, a 270mm instrument was completed in 1938. This larger telescope was destroyed in the bombing of Kutter's Munich property in World War II. In 1946 Kutter went on to make a third instrument, the "Else Kutter Medial". This all-spherical, 150mm instrument was based on Schupmann's article cited earlier. The curves were computed using Schupmann's, as Kutter put it, "unfashionable and nonconformist mathematical style". This last instrument (see Figure 12.4) was the first to use a single-element corrector thus setting the course for nearly all future interpretations.

To end the discussion of German-made Schupmann telescopes we must in-

Fig. 1.7 *Philipp Johann Heinrich Fauth's drawing of the Hyginus Rille (upper left) from his Der Mond und Hörbigers Welteislehre (1925) is typical of his meticulous depictions of the lunar surface. At the lower right Fauth poses in front of the dome that housed his 15.1-inch Schupmann medial (Photo courtesy José Olivarez).*

clude a giant brachymedial by the amateur astronomer, Edwin Rolf. This 700mm aperture *f*/19.4 instrument was completed in 1953, and is presently located in Rathenow and in use there (see Figure 9.28). A notable feature of the Rathenow instrument is a unique finder that uses the optically obstructed center of the objective as the first focusing stage to a down-stream located second achromatizing positive doublet lens, followed immediately by a diagonal mirror which sends the now low-power wide field to focus conveniently beside the main instrument's focus. This instrument is further detailed in Section 9.9

It is interesting to note that Schupmann's dream was to develop truly huge refractors capable of competing with reflectors of the era. But instead, his invention provides the amateur with a design perfectly suited for planetary observation and serious double star measurement, almost always best accomplished with instruments of comparatively tiny, 250–350mm apertures. The relatively recent introduction of the CCD makes the Schupmann's high definition and complete color correction even more attractive for the serious amateur astronomer. The author can do no better than quote Anton Kutter's summary of the instrument:

> ... the Medial is the specific scope for the finest definition and therefore the specific high power telescope.

Kutter has chronicled the development of telescopes from the earliest time ending with Schupmann's development of the single element corrector where he comments questioningly: "UTOPIA?" Only time will tell!

Thus ends the development and implementation of the Schupmann medial in the inventor's homeland, apparently to the present time, where his great medial principle went largely unnoticed. To quote Kutter: "Optical industries and professional optical analysts certainly acknowledged his work, but businessmen preferred to sell a big OG with two lenses than a similar one with only a single lens." We must turn to the United States to continue this account.

1.2 Developments in the USA

The U.S. patent granted Schupmann the rights to two types of medials, the first is called a brachymedial (shortened or broken medial). This design interposes the back reflecting meniscus lens corrector well before the objective's focus thus returning the converging cone back up the tube where it is intercepted by a flat mirror and brought to focus at the side of the tube as with a Newtonian. Because the diagonal mirror, in Schupmann's design, is located well down the converging beam from the objective it geometrically represents a larger obstruction than its size would first indicate. The otherwise perfect on-axis image suffers lateral color as off-axis images are examined. This can be corrected with a cemented three-element field group located a short distance before the eyepiece. Chapter 9 details Schupmann's brachymedial where various designs are presented, including a perfectly corrected modern version of the brachymedial.

The second type, the medial—meaning "somewhere between" (refractor and reflector in this case)—is the form of greatest interest to those wishing to obtain a high power unobstructed telescope. Fortunately, James Baker while at Harvard College Observatory recognized the optical correctness and possibilities of the medial form, and set out to clarify this greatly misunderstood design with an article in the *The Astronomical Journal*[8] in which he described an interesting design

[8] Baker, James G., "The Catadioptric Refractor", *Astronomical Journal*, Vol. 59, No. 2, March 1954, pp. 74–83.

for a 29-inch aperture medial to be used as a combined solar and planetary telescope. Baker concluded that the Schupmann is, by its very nature. (with the simple addition of an occulting disc, Lyot stop and interference filter), a solar coronagraph and prominence telescope. This characteristic of the design apparently escaped the notice of all previous investigators. It is the coronagraphic nature of the design, above all others, that sets the medial apart from other white light telescopes. Baker also showed in a *Sky and Telescope* article[9] how the Schupmann principle can be employed to correct the secondary color of the giant refractors still in use today. The author knows of no case where this has been done and, unbelievably, our great refractors remain plagued by intolerable secondary color!

Another article[10] by Joseph Rush published in *Scientific American,* told the story of the Schupmann in a very readable, accurate and descriptive style. Rush included enough "how-to" information for the advanced ATM to build a very nice 4-inch "in line" system.

Going back in time we do find that Schupmann's U.S. patent stirred some interest in the amateur community leading to the first known US interpretations by Hellman (1941), and Dwight and Gorden (1947) and one in Canada by McClung (1942). These instruments were of the brachyt or brachymedial type and were only experimental in nature (see Chapter 9). Still no US medials other than these three Brachy's of limited success had surfaced.

It is a private music teacher and amateur telescope maker named Chester Cook (1892–1963) that we must thank for the first successful medial in America. Cook was inspired by Baker's enthusiasm for the Schupmann and with his design help began work on an 8-inch *f*/15 employing a ⅓-diameter "Mangin" corrector. Completed in 1952, this telescope was a true medial employing a field lens at the first focus; however, this is where the similarity with the German designs ended. Rather than the usual "elbow" prism to turn the rays 90 degrees, Cook took a bold step and positioned his corrector on the center line of the objective thus introducing the "in line system". Although this leads to a longer tube, the balance on the usual tall refractor pier is actually improved as there is plenty of floor distance or clearance for the corrector housing. And, due to the tubes, a lower center of gravity means that the eyepiece can be somewhat lower than the conventional refractor. All efforts to find a photograph of this beautiful instrument have reached a dead end, made all the more poignant for the author by forgetting

Fig. 1.8 *Chester Cook. Photo courtesy Bert Willard and the Springfield Telescope Makers' archives.*

[9] Editors Report, American Astronomers Report. "Achromatic Correction for Refractors", *Sky and Telescope*, February 1951, p. 93.

[10] Rush, Joseph H., "The Amateur Scientist Department", *Scientific American*, May 1958, pp. 130–138.

Fig. 1.9 *Ed Olson 1907–1998 (left) and (right) Clifton Ashcraft talking with Ed at the 1969 Stellafane Convention.*

his camera on a visit with Chester at his backyard observatory shortly before he passed away. Apparently the optical parts were lost in the settlement of his estate.

Soon after Cook's medial was complete, Richard Dunn of Sacramento Peak Observatory and Jim Gagon of the Boston ATMs, working to a Baker design, completed the second US medial and then the world's largest, a 16-inch *f*/20 coronagraphic instrument. This somewhat complex version was built for Harvard College Observatory and was intended to replace the historic15-inch great refractor. Thankfully, it was instead mounted at Sac Peak. The instrument never went beyond the on-mount testing[11,12] stage as it apparently had wave-front error residuals associated with the toroidal figure on the mirrored surface of the corrector, thus the instrument was soon dismantled. (Toroidal figuring, favored by Baker on theoretical grounds, is an alternative method of correcting the astigmatism arising from this particular design's unusually large corrector tilt). Eleven years later, Joe Rush and George Schnable completed a complex16-inch Schupmann for coronal studies at Climax Peak, Colorado. This instrument, described in an *Applied Optics* article,[13] was mounted on an equatorial spar, but is not operating as a Schupmann at the present time.

A special moment in the medial's development came in 1952 when Edwin Olson read Baker's manuscript for the forthcoming *AJ* article. Being an advanced amateur with long experience with refractors, Olson easily caught the Schupmann bug and saw its potential to provide the amateur with the perfect planetary telescope. Baker put Olson in touch with Cook who kindly supplied a design prescription that Olson needed to start a medial of his own. Olson took the Cook design and modified the corrector to be the same diameter as the objective, the so-called

[11] F.Q. Orrall, Thesis, Harvard University, 1954.

[12] R. B. Dunn, *High Resolution Solar Observations*, Sacramento Peak Observatory, AFCRL, Sunspot, New Mexico, USA, Space Research XII - Akademie-Verlag, Berlin, 1972.

[13] Rush, Joseph H. and Schnable, George H., "High Altitude Observatory's New Coronagraph and Spectrograph", *Applied Optics*, Vol. 3, No. 12, pp. 1347–1352.

full-size system. The design was scaled to a 6-inch aperture. The telescope was unusual in that it incorporated a field mirror, suggested earlier by Baker, rather than the traditional field lens/prism arrangement. The completion of this instrument represented the first practical design for amateur construction. After completion and field tests this medial[14,15] (see photo Figure 12.8) was loaned to the author for further evaluation. Its performance was remarkable in every respect, which led in 1966 to the formation of the Schupmann Club[16] whose sole purpose was to develop the Schupmann for amateur construction and acquaint the amateur world with the possibilities of the medial telescope. The most noteworthy development was a discovery by club member Clifton Ashcraft. Ashcraft's ray trace investigation[19] revealed that an all-spherical solution existed when the corrector diameter equaled 54.4% of the objective diameter. This discovery led to what we call the Super Schupmann and is the basis for the "Standard Design" detailed in Chapters 9 and 10. The club was only partially successful. Although about five instruments were completed, there still seemed to be a reluctance for the advanced TN to take a strong interest in the design. Photos and descriptions of telescopes built during the time the club was active, along with those of later instruments, can be found in Chapter 12.

It was more than 80 years after being made public that Ludwig Schupmann's design for a perfectly color corrected refractor was given a measure of acceptance when Mike Mattei of the Boston ATMs received first place in the optical excellence competition for his 6-inch *f*/15 planetary medial[17,18,19] at the 1983 annual Stellafane convention on Breezy Hill, Springfield Vermont.

At or about this time the Schupmann Club was dissolved and the author's ATM activity was put on hold in order to write a booklet called *Amateur Construction of Schupmann Medial Telescopes.*[20] It was felt that putting everything learned about making and using the medial in a single, handy, how-to form would best serve the amateur community. It took about a year to bring the manuscript into shape and with a great effort the booklet was printed just in time for the 1984 Stellafane convention.

It was 1990 when, finally, word of a newly completed Schupmann reached the author. This 7-inch *f*/10 was built by California amateur, Gerry Logan. His beautiful instrument is shown in Figure 12.10. Gerry built the all-spherical Super

---

[14] Olson, Edwin A., "Schupmann for Amateurs," *The Maksutov Club Circular*, No. 119, September 1964, pp. 15–19.

[15] Cox, Robert E., "Amateur Astronomers" conducted by Robert E. Cox, "Stellafane 1964 Highlights," *Sky and Telescope*, October 1964, pp. 204–205.

[16] Daley, James A. Jr., "The Schupmann Club," *Sky and Telescope*, March 1983, p. 278.

[17] Overbye, Dennis, "A Night on Breezy Hill," *Discover Magazine*, October 1983, pp. 84–87.

[18] McGregor, Doug, "Amateur Astronomers" conducted by J.W. Briggs, "Stellafane'83," *Sky and Telescope*, November 1983, pp. 444–447.

[19] Berry, Richard, "Of Stellafane Optics and the Art of Deep Sky Observing," *Deep Sky Monthly*, Winter 1983, pp.33-34.

[20] Daley, James A. Jr., *Amateur Construction of Schupmann Medial Telescopes*, privately published, 1984, 63 pages.

Schupmann described in the construction booklet. Gerry is a top notch mirror maker and a highly talented woodworker which all came through in his work on this instrument. He received a merit award at the Riverside convention for his efforts.

Shortly after completion of the Logan instrument, Scott Milligan of the Springfield Telescope Makers informed the author of his plans to head a club effort to construct a 13-inch *f*/10 Super Schupmann, again based on the prescription in the booklet. This club project was monumental in scope involving as it did over 40 dedicated and highly talented club members and included building the McGregor Observatory (at Stellafane) and telescope mount as well. The potential of the Schupmann telescope has been fully realized with this instrument. In addition to the pure joy of recreational cosmic touring, the Stellafane Schupmann is also employed in CCD photometry of eclipsing binaries, which has been extended to include accurate CCD photometry of close visual binaries. This work requires the highest resolution to obtain, without cross-contamination, scientifically valuable data in BVRI bands. See Figure 12.1 and Chapter 13 for more on this instrument.

Five years after the completion of the McGregor Observatory the author received a wonderful letter from Gerry Logan telling of his latest Schupmann, a 10-inch *f*/9.75 folded Super Schupmann. This represents the world's first folded medial. It's of enormous practical interest because, if successful, it will make possible a very short instrument. For years Schupmann builders have talked of the bad effects of adding an additional reflecting surface (flat). Our fears were unfounded, as Gerry reports no reduction in contrast and that the idea is perfectly practical. As Figure 12.6 shows, the eyepiece is in a sort-of Newtonian viewing location. Even allowing for the collapsible dew and body-heat shield, the instrument length is only slightly over 60 inches. Gerry uses it mostly in the study of tiny planetaries.

1.3 Back to Europe

Recently, news of a giant medial reached the author. This professional instrument[21] became operational in May 2002 and is by far the most significant earth-based solar telescope in existence. Located on La Palma in the Canary Islands at an altitude of 2400 meters, the instrument is operated by the Institute for Solar Physics of the Royal Swedish Academy of Sciences (see Figure 1.10). This 0.97-meter clear aperture medial operates at *f*/21 and utilizes the field mirror configuration. Because of the solar heat loading, optical grade quartz was determined to be the best choice for the objective. The corrector is configured to operate in double-pass and is also quartz with the retroreflection performed by a separated concave mirror of Zerodur. To suppress harmful internal air currents, the entire optical path is evacuated with the objective serving as the vacuum (entrance) window. A turret atop the tower directs the light from the objective, first to a 1.4-meter Zerodur elevation flat and thence to an identical azimuth flat where the beam is

---

[21] Scharmer, Göran B., Bejelksjö, Klas, Korhonen, Tapio, Lindberg, Bo, Petterson, Bertil, "The 1-meter Swedish solar telescope," *Proceedings of SPIE,* Vol. 4853, 2003.

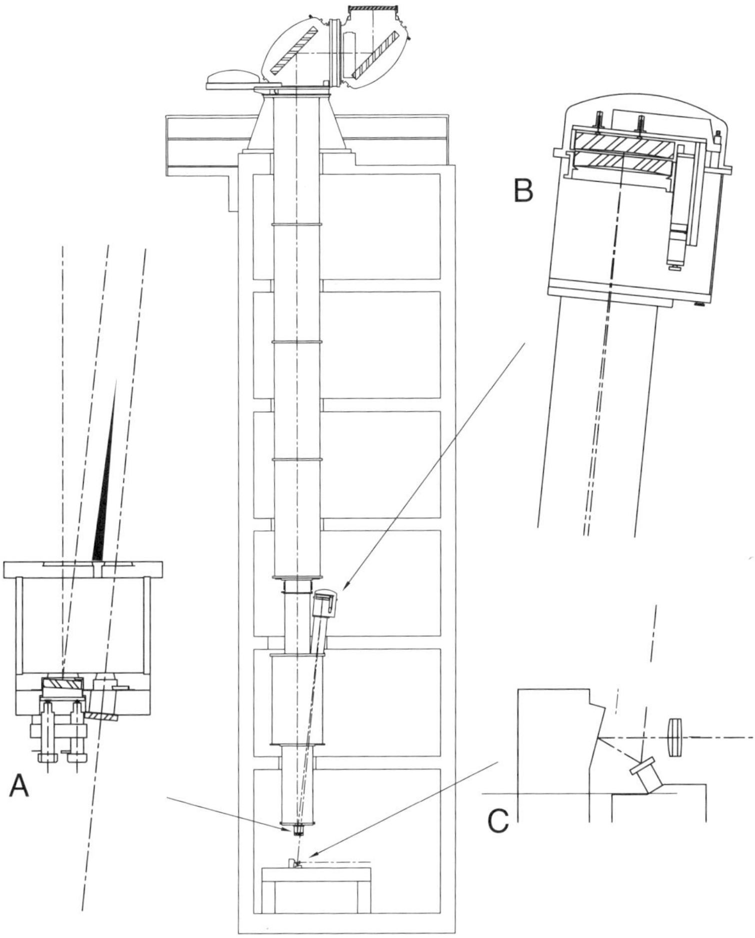

Fig. 1.10 *Schematic view of the 1-meter Swedish Solar Telescope (NSST). The quartz objective lens is the skyward facing element of the two-mirror Alt-AZ turret. Fine control of the field mirror tilt, shown in detailed view A, is performed with two orthogonal piezo actuators. A water-cooled aperture plate and special stray light baffle along with the beam exit port lens are also shown in A. The two-part corrector is shown in B. It is Schupmann's second type (see Figure 9.2). Detailed view C shows the optical bench where a final pupil, formed by the exit port lens, is imaged on a deformable mirror. The bench also holds a fast steering mirror, wavefront sensor, final imaging lens and science CCDs. Drawing courtesy of Göran B. Scharmer, Institute for Solar Physics of the Royal Swedish Academy of Sciences.*

sent vertically downward to the first focus at the field mirror. The slightly tilted field mirror reflects the now diverging cone obliquely upward to the corrector. The corrector, operating as a 1-to-1 relay, sends the beam back to focus close beside the field mirror. The fully color corrected beam exits through a field lens (which seals the vacuum chamber) forming a second downstream pupil on the deformable mirror located on an optical bench. Other bench optics include a high quality final imaging relay lens, a wave-front sensor, and CCD cameras. A unique property of the medial design, which was recognized and utilized by amateurs long ago, allows the cancellation of atmospheric dispersion by appropriately tilting the field mirror which slightly translates the pupil image at the corrector, thus introducing apposing dispersion almost perfectly offsetting the atmosphere's. The Swedish instrument utilizes piezoelectric drivers to control the tip-tilt of the field mirror for this function. One novel idea conceived and implemented by the Swedish designers is their method of focusing by means of axial motion of the lens element of the corrector, apparently without harmful effects.

Recent white-light solar images obtained by the Swedish instrument are of extremely high resolution and have revealed sunspot penumbra filaments to a level of ultimate detail. The age-old speculation on the nature of faculae as "floating hot gas clouds," etc., has been disproved and the exact cause of their white appearance finally determined[22] with this medial's beautiful image quality. This image quality is further demonstrated in the July 2004 *National Geographic* magazine which features the latest developments in solar research. Most recently[23] the instrument's unprecedented resolution has helped to arrive at a new theory of spicule generation by magnetic flux tubes which pierce the Sun's cooler upper layer—amazing!

After 102 years of near neglect the Schupmann is no longer an optical curiosity, instead it has come of age! How would Schupmann react now: tears of joy for sure.

At the time of this writing, construction of other medials is underway. One of these was to be an 8-inch planetary brachymedial by Klaas Honders of the Netherlands, but his untimely death will regrettably leave it unfinished. Klaas' designs appear in Chapter 9.

The future scientific use of the medial probably lies with the serious amateur astronomer doing specialized measurements. For the avid TN nothing compares with the joys of completing a top-notch medial!

1.4 List of All Known Schupmanns to Date

Keeping an up-to-date list of completed Schupmanns has been a lot easier than the work of an 8-day clock winder. Only when the exact dimensional details of some of the older instruments are researched do difficulties arise. So, dear reader, look upon this list somewhat provisionally as some historical information is conflicting (fortunately only in tiny millimeters), so, overall, the list is pretty reliable.

[22] *Sky and Telescope,* February 2003, page 16.

[23] *Sky and Telescope,* February 2003, page 16.

Table 1.1 List of All Known Schupmann Telescopes

Maker	Year-Country	Aperture and F#	Type	Status
Schupmann	1893-Germany	127mm-*f*/13 (est)	Brachymedia	Unknown
Schupmann	1897-Germany	127mm-*f*/12.8	*Prism/Field Lens	Unknown
Reinfelder and Hertel	1900-Germany	333mm-*f*/15.0	Prism/Field Lens	Destroyed WW1
G and S Merz	1911-Germany	385mm-*f*/10.6	Prism/ Field lens	Destroyed
G and S Merz	1921-Germany	323mm-*f*/11.5	Prism/ Field Lens	In use
Kutter	1936-Germany	122mm-*f*/10	Prism/ Field Lens	Unknown
Kutter	1938-Germany	270mm-*f*/10	Prism/ Field Lens	Destroyed WW2
Hellman	1941-USA	108mm-*f*/13	**Brachymedial	Unknown
McClung	1942-Canada	83mm-*f*13.8	Unknown type	Unknown
Kutter	1946-Germany	150mm-*f*/11.3	Flat Mirror /Lens	Unknown
Dwight/Gordon	1947-USA	155mm-*f*/25	Brachymedial	Unknown
Cook	1952-USA	203mm-*f*/15	Field Lens in line	Lost
Dunn/Gagen	~1953-USA	406mm-*f*/15	White Light Coronagraph	Stored at Sac Peak
Rolf	1953-Germany	700mm-*f*/19.4	Brachymedial	In use
Olson	1963-USA	152mm-*f*/12	Field Mirror	Donated/stored
Rush/Schnable	1964-USA	406mm-*f*/20	Field Lens, Corona-graph-Spectrograph	Unknown
Daley	1969-USA	190mm-*f*/13.5	Field Lens	Reconfigured 1988
Olson	1970-USA	108mm-*f*/11.3	Field Mirror	Donated/lost
Prugnarola	1970-USA	190mm-*f*/13.5	Field Mirror	Stored
Olson/Daley	1973-USA	108mm-*f*/11.3	Field Mirror	Donated/In use
Mattei	1983-USA	152mm-*f*/15	Field Mirror	In use
Daley	1984-USA	152mm-*f*/15	Field Mirror	Donated/In use
Daley	1988-USA	190mm-*f*/13.5	Field Mirror	Optics stored
Logan	1990-USA	178mm-*f*/10	Field Mirror	In use
Daley	1990-USA	228mm-*f*/11.1	Field Lens	Stored
Bray	1990-USA	150 mm-*f*/12	Unknown	Star Tested, Unfinished
Milligan	Stellafane-1995, USA	336mm-*f*/10	Field Mirror	In use
Logan	2000-USA	260mm-*f*/9.75	Folded, Field Mirror	In use
Mattei	2001-USA	254mm-*f*/10	Field Mirror	Star Tested
Daley	2002-USA	230mm-*f*/11.1	Field Mirror	In use
NSST	2002-La Palma, Spain	970mm-*f*/21	Field Mirror	In use

* This configuration "Prism/ Field Lens" is Schupmann's familiar 90-degree "elbow" arrangement. The field lens is usually ground-on or cemented-to one surface of the prism face.

** The "brachymedial", or its German synonym "Verkürzter Refraktor" (meaning shortened refractor), lacks an intermediate focus and associated field element, thus is more difficult to correct for lateral color (usually requiring a cemented three-element lens near the focus). However, modern interpretations by K. Honders, using large correctors, show remarkable performance in ray trace studies and display near astrographic fields of view.

Chapter 2
The Schupmann and Amateur Research

2.1 General Philosophy

As curiosity is the springboard, measurement is the stuff of science. Our measures may bolster or disprove a theory or lead to development of a new theory. To be an amateur astronomer one must, above all else, measure and measure accurately, just as the professional does.[1] Good measurements require carefully considered instrumental hardware, usually designed specifically for the project and often built, or at least implemented by, the amateur alone. The amateur stellar scientist has no time or desire to custom-knurl his focus knob or hand rub the varnish of the tube. Instead he dedicates all his design skill and construction energy to the details of the measurement task. After a long period of astronomical measurements, followed by peer review, the amateur publishes his data and possible findings in an appropriate journal. Astrophysicists working in the amateur's field will read these reports and use the data beneficially. Observations which lie in a notebook never to see the light of day, do no one any good. Sometimes the amateur scientist actually builds his own telescope. The author dedicates this chapter to those very few.

Considering the above, one may wonder, in an age of giant telescopes, if there is still a place for the small amateur instrument in useful scientific study. The answer is a resounding yes, and we need only look at the ongoing work of amateurs cooperating with professional astronomers in the fields of variable stars, asteroids and binary stars to see the successful use of the small telescope. The recent development of the CCD camera has given the small instrument new life and the user a powerful tool to carry out these, and other research programs, to a level that could only be dreamed of 20 years ago. These precise imaging arrays allow photometric and astrometric measurements of the highest accuracy, thus making the work of the amateur and professional nearly indistinguishable in many areas of research and discovery.

Other than a few diehard double star observers using filar micrometers, and the dedicated variable star observer who has yet to switch to a CCD; the scientific

[1] C. Starken, *The Study of Variable Stars with Small Telescopes*, Edited by John R. Percy, Cambridge University Press. 1986, "Variable Stars as Targets for Long-Term Coordinated Multisite Photometry", pp. 165–179.

role of visual astronomy has essentially ended. We find instead, thousands of hobbyists gazing through their scopes purely for pleasure and the usual bragging rights. Yes, now and then a comet or supernova is discovered visually but even this area of astronomy is being rapidly replaced by automatic search "machines". The pretty CCD pictures of celestial objects that fill the popular astronomy magazines sell cameras and telescopes but cannot be considered amateur research or even amateur astronomy for that matter, they are simply the product of a very enjoyable, and beautiful, technical art form. The author still can't resist, once in a while, installing the focal reducer in his 9-inch medial and playing astrophotographer. May the hobby happily continue. He also enjoys peering into the eyepiece at a beautiful globular cluster or Jupiter's incredible cloud belts, but these are brief interludes between long periods of making measurements, both astrometric and photometric, of double stars with a CCD at the focus of his 9-inch Schupmann medial. Each clear night the seeing and the scope's optical performance are visually evaluated on a twilight star but soon the CCD is installed and the night's fascinating work begun.

We will now address the Schupmann medial's suitability for amateur astronomical research.

2.2 Astrometric Considerations

2.2.1 Aperture and Focal length

The amateur astrometrist's first step in choosing a dream telescope is deciding on the aperture and focal ratio to satisfy the requirements of their observing program. Usually the aperture is determined by four considerations:

1. Stellar magnitude of program objects; i.e., asteroids, double stars, etc. Using an affordable CCD we can scale from the performance of a typical 8-inch *f*/4 Newtonian where 15th magnitude stars are solidly imaged in ~60 seconds. Valuable asteroid work is easily performed with such a configuration, although the field of view is limited by coma without special correctors. For double star measurements an instrument must be operated at a much higher f# with *f*/27 (independent of aperture) being about right to reach the necessary imaging resolution. Here a basic 100-inch focal length medial system is usually amplified with a 2 or 3x Barlow. Due to the drive precision needed for this type of work the exposures are typically limited to about 20 seconds, and the images of useful "density" will fall in the range of 13th magnitude.
2. Resolution requirements, in view of atmospheric blurring, are usually satisfied with apertures in the range of 7 to 12 inches.
3. The cost of the entire observatory is a serious issue and it is here that the aperture can heavily impact one's scientific wishes. You can count on the cost going up as the cube of the aperture!
4. The challenge of making optics for ever larger instruments of truly dif-

fraction-limited performance imposes limitations for each builder in many varied ways. It goes without saying that amateur-made telescopes of "astrometric" quality (properly mounted and in an observatory) of over 12 inches are practically nonexistent!

2.2.2 Field of View

Having decided on the aperture and focal length, the next optical concern is the choice of which instrument design will provide satisfactory image quality over the desired field of view. The largest affordable CCD usually determines the field of view, in practice. The Schupmann medial telescope is one of many unobstructed systems employing tilted elements. This characteristic of the design limits the field of diffraction-limited images to a smaller value than, for instance, a comparable apochromatic refractor. Of course, each medial configuration should be separately analyzed to determine the useful field. In this respect designs with low or equal objective-to-corrector diameter ratios show the largest astrometric-quality fields, typically ¼ degree or greater. Although the Schupmann medial cannot be considered a wide-field instrument it has many desirable astrometric properties relating to the measurement of double stars where fields of a few hundred arcseconds encompass just about all (interesting) binary separations. The medial's image quality over these small fields is truly diffraction limited. Larger astrometric fields are obtained with new versions of Schupmann's shortened design called the brachymedial. This untilted form avoids an intermediate focus as the corrector is positioned well before focus thus folding the objective's converging beam back upon itself where, just before focus, a diagonal mirror reflects the beam outside the tube. Modern interpretations of this configuration by the late Klaas Honders of the Netherlands are described in Chapter 9. Utilizing a large format CCD, near diffraction-limited astrometric, or more to the point, astrographic quality fields of over 3 degrees are obtainable with one of Honders' designs.

2.2.3 Field Distortion

Another important astrometric consideration is the level of field distortion. Most tilted-element telescopes are afflicted with distortion, usually anamorphic in nature. The Schupmann fares very well indeed in this regard as the distortion is of a very weak elliptical form. As with field aberrations lower distortion is achieved with large correctors. Additionally, making the corrector as thin as possible can further reduce this aberration by causing the entering and emerging rays to more nearly coincide. The ray trace work performed by Bert Willard (see Figure 2.1) shows that the author's 9-inch medial's worst distortion, edge-to-edge of the 100 arcsecond field studied, is only 0.01 arcsecond (0.01%). (It should be noted that this field is centered on the instrument's full ½-degree FOV). This is a factor of at least four smaller than the typical measurement errors for a doubles components lying at opposite borders of the field studied. The distortion terms can be applied in the data reduction process for the most demanding studies, as they remain a permanent characteristic of the instrument. It is often observed that absolute errors in

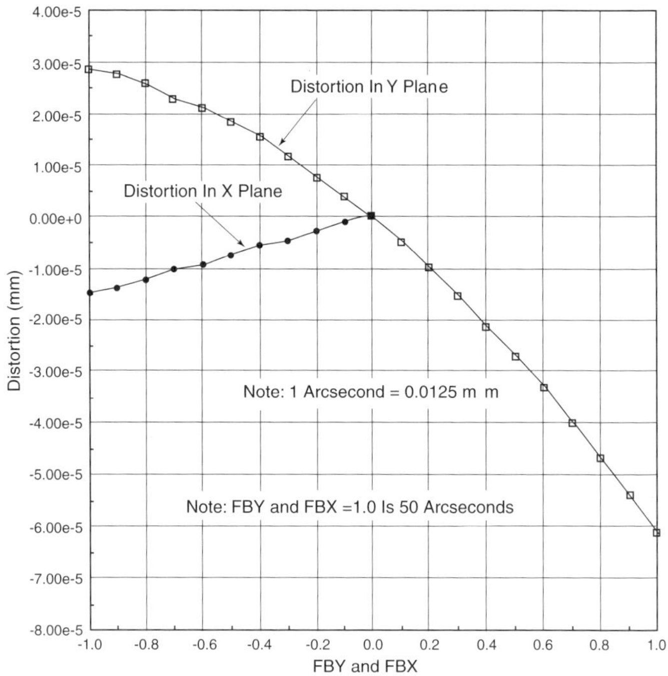

Fig. 2.1 *Bert Willard's field distortion analysis of the author's 9-inch f/11.1 medial.*

determining the separation of wide doubles are somewhat greater than for moderately close pairs. This is due, primarily, to imprecise knowledge of the focal length (addressed later) and to a lesser degree on uncorrelated seeing. Typically, with a well-calibrated astrometric sensor, the errors (in fractions of an arcsecond) are pretty much the same for all separations, leading to the greatest *percentage* errors for the closest pairs. The more important doubles for amateur research have separations in the range of 5 to 30 arcseconds. At these separations the worst-case distortion is smaller by a factor of about 30 than the measurement uncertainty. As an example, by averaging the measures obtained from 12 CCD frames of an equal 8th magnitude double star of ~20 arcsecond separation, the author finds a probable measurement error of ± 0.02 arcsecond (±0.1%) when using his 9-inch medial working at an effective (Barlow amplified) focal length of 301.76 inches. This high level of accuracy is achieved only with excellent seeing, high signal-to-noise ratio (s/n) star images and careful use of astrometric programs. If the field of view is pushed to its limits for other types of astrometric work, such as asteroid positions, it's a good idea to arrange the CCD's rectangular field such that the short dimension is in the direction of the greatest distortion for best overall field performance. As a final comment it should be noted that for specialized work, such as binary star measurement, the element tilts could be greatly reduced for a smaller overall FOV which leads to a virtually distortion-free system.

2.2.4 Field Curvature and Field Tilt

Let's consider the field curvature or curvature of the focal surface first. Here again, as with other field defects, the most favorable values for curvature are achieved with large correctors. The sign of the field curvature is positive in the medial which means it is backward-curving (concave to the eyepiece). As shown by Schupmann and confirmed by Kutter, the field curvature is deformed toroidally, however, these deformations (in the modern medial) do not depart from a best-fit sphere enough to be significant. The radius of the field curvature varies linearly with corrector diameter for any given objective. Again, taking the author's 9-inch (½-size corrector) as an example, the curvature of field is so shallow that no defocus will be observed across the typical CCD chip. The shallowest field curvature is obtained with objective-to-corrector diameter ratios of one, the so-called "full size" systems. As an example, for a full size corrector in an 8-inch *f*/20 field mirror system we find the curvature of field radius to be 867 mm (this is the Petzval radius and because of the medial's dominant aberration, astigmatism, can be used as an approximation of the best-focus surface) giving a defocus across a 1-cm CCD chip of 14 microns, or a bit over one-half thousandth of an inch—totally undetectable at *f*/20! On the other hand, medials with tiny correctors exhibit a rather strong curvature of field. An 8-inch *f*/14 example, utilizing a 22% diameter corrector, will have a curvature of field radius of 68 mm, leading to a defocus of 0.18 mm when sharply focusing on a 1-cm diameter field center. This value of defocus will, theoretically, exceed the depth of focus tolerance for the typical focal ratios employed in Schupmann telescopes. One, in practice, would focus for the 70% field zone thus dividing the harmfulness by two. With a weak positive lens near the final focus, the toroidally deformed field curvature could be greatly reduced but this is not easily implemented with a CCD. Also such a strategy will introduce lateral color and other aberrations far outweighing any intended gains. As a practical note the author has yet to detect curvature of field in any of the many medials he has constructed, either with a CCD or visually.

The field tilt is typically equal to approximately twice the corrector tilt and thus you see why the larger correctors with their smaller tilts are again favored. Figure 4.6 is taken from page 99 of *Die Medial-Fernrohre* and shows, in a highly exaggerated way, the direction of tilt with respect to the corrector tilt and hints at the slightly warped or hooked curvature of field. The Standard Medial described in Chapters 5 and 10 has about 0.3-degree field tilt, but if necessary, this is easily compensated for with a simple matching tilt of the CCD head.

2.2.5 Focal Length Stability

One important instrument characteristic that impacts astrometry is the stability of the focal length over its operational temperature range. For wide-field CCD astrometry, most computer programs allow the operator to recalculate the focal length for each CCD frame, as there are well-calibrated positions for certain stars in the field. The situation is very different with the tiny fields (long effective focal lengths) needed to achieve the highest accuracy in double star measures. Many

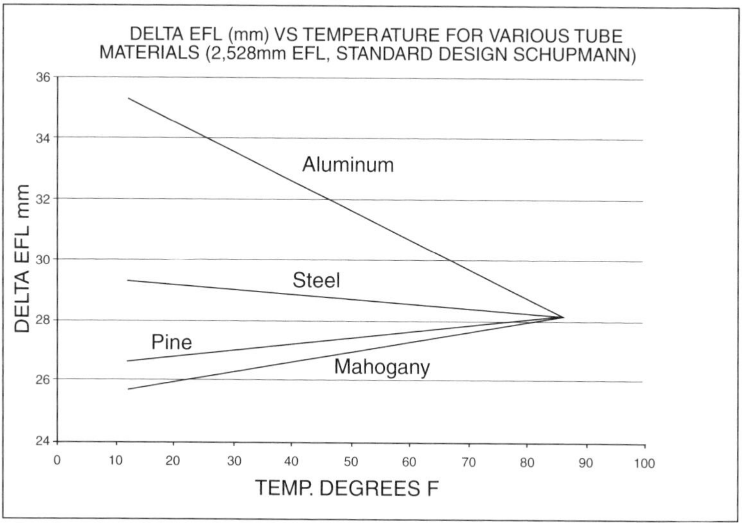

Fig. 2.2 *The author's analysis of focal length stability with various tube materials as applied to the Standard Design.*

double star observers choose to determine the system EFL independent of even "well known pairs" and instead use single star-drift (transit) methods to calibrate the filar micrometer's screw constant or, in the case of the CCD, system EFL. Obviously, night-to-night knowledge of the telescope's focal length is of extreme importance. Ideally one would design the optical system's EFL to be totally immune to temperature changes.

Now, with all that said, let's look at this problem as it applies to the Schupmann medial. Temperature variations change the index of refraction (dn/dt) and optical curvature (coefficient of expansion) of the medial's lenses, slightly varying their focal length. Luckily the thermal characteristics of BK7 glass tend to nearly cancel resulting in a change in focal length for the objective of only 1 part in 22,000 over a temperature range of 20 degrees centigrade. The corrector is only slightly more sensitive because of its lens/mirror nature. Unlike a normal refractor, the Schupmann's tube acts as a very long lens spacer and the material must be chosen such that the thermal expansion properties minimize EFL variations. Figure 2.2 shows the results of a ray trace analysis of the standard medial of 2528.0 mm (~100-inch) focal length employing different tube materials. (It is the last two numbers in mm before the decimal point that are shown in the graph.) Note how the change in focal length can be *positive* or *negative* depending on the material chosen. The temperature range for the study is consistent with the author's observing conditions. It is clear that the temperature effects are measurable, however one must put the question of focal length stability in perspective as it pertains to astrometric measures.

The important parameter is the temperature-induced focal length shift as a *percentage* of the focal length. A one percent change in focal length translates, if

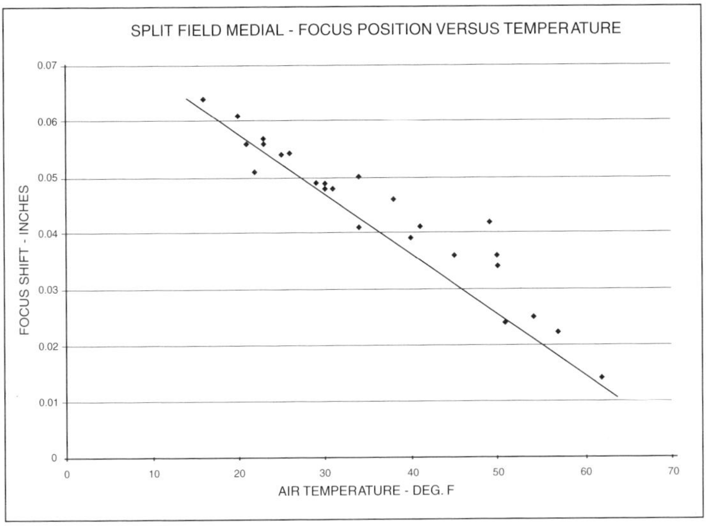

Fig. 2.3 *Comparison of focal position data (crosses) with computer ray trace prediction (line) for the author's de-commissioned split-field medial.*

unaccounted for, to a one percent error in measurement. Determining the focal length to this accuracy is quite a task to begin with and the avid astrometrist will go to incredible effort to obtain a probable error of ±0.1% in the determination of his instrument's (single temperature) focal length. As we saw in the discussion on distortion, knowledge of an instrument's focal length to this precision is more than adequate for solid, scientifically meaningful, binary star measures. Exact knowledge of focal length vs. temperature allows the astrometrist to apply tiny corrections to their results as long as the thermal settling characteristics introduced by the usual falling nighttime temperature are well understood. The author has studied the stability of the *focal position* of his earlier 8-inch *f*/14 split-field medial which is an in-line configuration employing a field lens. The measurements are referenced to the focus platform, which is bolted to the mahogany plywood tube. The plotted data points are shown in Figure 2.3. Focal position stability is not a direct measure of focal length stability but does indicate that things are not going wild. Included in the graph is a ray trace result (solid line), which shows remarkable agreement with experimental data. These results show the EFL changing by only 1 part in 1375 over quite a temperature range. This observation is confirmed, not only by the ray trace, but also by the freedom of systematic winter/summer O-C ("observed minus calculated") errors in the author's double star measures. It can be concluded that the focal length stability of the split-field in-line system as well as the field mirror type Schupmann, competes favorably with the normal achromatic objective traditionally used in astrometry.

It may be possible for the real stickler to "athermalize" their medial by the combined effect of the tube and internal nylon and steel spacer rods acting on the relay (corrector) element to exactly maintain the effective focal length over huge temperature extremes. One would need to take care to get the sign of the compensation correct. The spacer rods could take the form of return rods similar to "gridiron" compensating pendulum rods used in old precision clocks. Here (3) nylon rods would be anchored to the corrector cell mounting platform and extend toward the field element a calculated distance and be returned and attached to the cell itself with steel rods, all nicely guided in close fitting bushings. To reverse the sign simply interchange the rods. The internal location of these compensating elements allows them to, more or less, follow the temperature of the elements themselves, thus pretty much preventing overshoot of this part of the thermal compensation. Field mirror and field lens systems will behave differently depending on the layout. Plywood has a very low coefficient of thermal expansion and can be moisture-stabilized with epoxy coatings. The cross grain layers' greater thermal expansion is largely resisted by the grain layers that are parallel to the tube's length. Any compensation system imagined which includes the tube will undoubtedly slightly overshoot during the initial exposure to the night sky. Covering the tube with aluminum foil can largely negate this overshoot. These tiny compensations are harmless to the color correction and image quality. Ray trace analysis reveals that the use of aluminum for the tube, in lieu of special compensation or mathematical correction, will exclude the medial as a top-notch astrometric telescope.

Referring again to Figure 2.2, steel is obviously a good choice for the tube but manufacturing difficulties and excess weight might be problems. Mahogany planking, on the other hand, is only slightly inferior providing a focal length stability better than one part in 1,000 over a temperature range of 40 degrees centigrade, thus well within our 0.1% astrometric requirement (one-half this when the focal length calibration is performed in the center of the expected temperature range). Walnut and chestnut will provide near perfect compensation but have a cost drawback that will become apparent when the builder visits the lumberyard! Epoxy coated pine plywood could be tried as it does show excellent thermal characteristics. In summary it can be seen that the Schupmann medial's EFL can be made very stable over a wide temperature range making it unnecessary to apply corrections to astrometric data.

2.2.6 Diffraction and Atmospheric Dispersion

Diffraction and atmospheric dispersion effects can bias the positional measures of stars. Diffraction can be troublesome when a large central obstruction and its associated shadow on the primary shift in the pupil as a function of field angle, thus slightly affecting the location of the photocenter. This is really only a problem with astrographic fields and is of minor concern in this discussion. Of course spider diffraction will alter the theoretical image as well. In Schupmann's medial configuration the unobstructed pupil remains unchanged over the useful FOV and is thus immune from such degradations. And, the Schupmann's unique optical properties provide a simple method of compensating for atmospheric dispersion (first order wise) for low elevation stars without the use of Risley prisms and the like. This is discussed in Section 4.2.4 and touched upon throughout this book.

2.2.7 Photometric Considerations

The great power of the CCD is very apparent in the field of stellar photometry. Other than for the exacting work of precise photometry of the close components of binary stars, the instrumental requirements are relaxed compared to astrometry. For variable stars, image quality is not as critical although the amateur photometrist would not be happy measuring images with huge coma flare! Color correction is, however, very important unless one is willing to refocus for each filter in the B, V, R, I bands. This refocus is usually required only for a traditional achromatic objective and is of no concern with the medial. The color response (transmission verses wavelength) of the telescope should be stable insuring that calibrations on standard stars are, in effect, semi-permanent. Color response permanence is best realized with sealed-tube telescopes, as in the Schupmann, which protects the reflecting surfaces from atmospheric degradation. While on this subject it should be mentioned that the ultraviolet wavelengths remain almost unobservable with refractive optics, especially when coupled with cheaper front-side illuminated CCDs. Instead of BK7 the builder could utilize the more expensive UBK7 or even better, optical grade quartz, which will substantially extend the transmission into the violet region. Care must be taken to prevent the slightest dewing of the objective in photometric observations, so an effective dew-shield should always be fitted. For less critical photometric observations an objective lens heater could be employed. Unlike the Schupmann medial's field limitations for combined photometry/astrometry, the FOV for variable star photometry can be much larger allowing, in many cases, the comparison star to be simultaneously framed when using a medium format CCD coupled with a focal reducer. The author has recorded many images through his medial using a focal reducer and a small format CCD camera giving a field of view of 9.3 x 7.0 minutes of arc, and their quality is perfectly suitable for photometry. Large format CCDs coupled with a focal reducer extend the useful field to over ½ degree. Incidentally, one feature of a proper photometric telescope is freedom from sky-flooding which would otherwise contaminate the data. The Standard Medial design is perfectly sky-baffled without the usual attendant vignetting. This relieves the sensitivity to alignment in the CCD flat-field correction process. Often the measurement of binary stars involves photometry in B, V, R, and I bands—for instance, in the determination of stellar association membership—and the author finds the Schupmann admirably suited to this dual role.

In summary, the medial design easily fulfills the requirements of moderately wide-field astrometry, and stellar photometry over substantially larger fields.

2.2.8 Planetary Research

The Schupmann provides nice, high-contrast, images of the planets. Again the CCD has all but replaced the eye, although some very limited visual planetary research (monitoring) can be performed. Data concerning Martian polar caps and the onset of dust storms are always welcome, although spacecraft observations are making this work redundant. Spotting the onset of Saturn's white spots would be

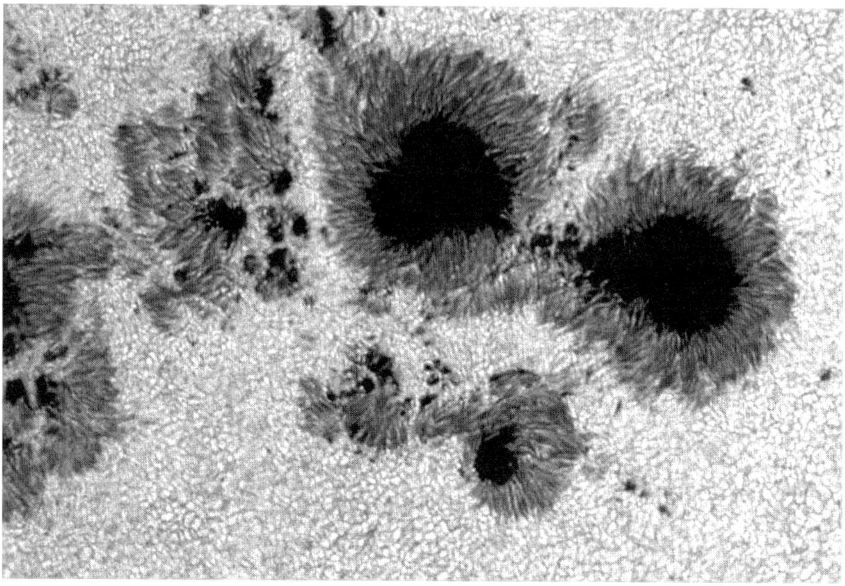

Fig. 2.4 *CCD V-band image of active region #0652 taken by the author with his 9-inch Schupmann medial on July 21, 2004. Details: 7-inch clear aperture Baader "AstroSolar" film filter centered over the objective, EFL = 278.82 inches, CCD = ST-7XE, Exposure = 0.12 seconds. Note the size compression of the granulation cells near the spots where the magnetic field strength is very high.*

useful for timely training of the Hubble Space Telescope on those features. Amateur planetary observations have been vastly superseded by those from spacecraft, so the visual observations of the future will be more in the nature of pleasant hours simply enjoying the view. Alas, the time has nearly passed for scientifically meaningful amateur contributions in this field.

2.2.9 Solar Studies

There can hardly be a telescope design better suited for solar research than the Schupmann medial. Unlike the planets, the Sun's moment to moment variations require the attention of both amateur and professional observers from stations spaced around the world. Presently the world's finest professional solar research telescope is a Schupmann medial located in La Palma Spain and is the chief instrument of The Institute of Solar Physics of the Royal Swedish Academy of Sciences (see Section 1.3). Operating at *f*/21 this 0.97-meter aperture Schupmann is very similar to the Standard Design detailed in this book. It utilizes a field mirror as in the Standard Design but has a separate reflecting element for the corrector. The corrector clear aperture is ¼ the diameter of the objective. The optical path is essentially vertical and can be filled with helium for reducing internal air currents. Or, for the best internal seeing the path is evacuated. Even though professional activity in this subject is intense, the fields for amateur research remain enormous—sunspot counts, flare spotting, study of prominence and sunspot structure (see Figure 2.4) and granulation lifetimes to name a few. And just think, you get a chance to sleep at night! This book covers many aspects of the medial design pertaining to solar work. It is notable that the only professional Schupmanns built in America were expressly designed for spectroscopic and white-light studies of the inner corona. The coronagraphic nature of Schupmann's medial configuration opens all sorts of experimental avenues. The serious amateur solar researcher may want to look closely at the medial's possibilities for their program.

Chapter 3
The Schupmann Principle and Practical Applications

3.1 The Schupmann Principle

We hear of the Schupmann principle in connection with many optical ideas, old and new. These ideas have in common a fundamental property that results when same-glass powered elements are so arranged that a color-free emergent beam results. Eyepieces have been designed by Berton Willard which embody the Schupmann principle and two of his examples built by the writer show perfect, color free images. It is the combination of identical, or nearly so, glass types that also provides the opportunity to achieve absolute achromatism in a telescopic optical system, and this property led Schupmann to his great invention, the medial telescope. If only the partial dispersions of crown and flint were perfectly matched, then Schupmann's ideas, if hit upon at all, would remain an academic curiosity as the harmful secondary color of our great refractors would all but vanish. Figure 3.1 shows how the medial telescope lies quite alone in an evolutionary corner.

To graphically demonstrate the Schupmann principle, as applied to his medial telescope, a sketch of the simplest arrangement is shown in Figure 3.2.

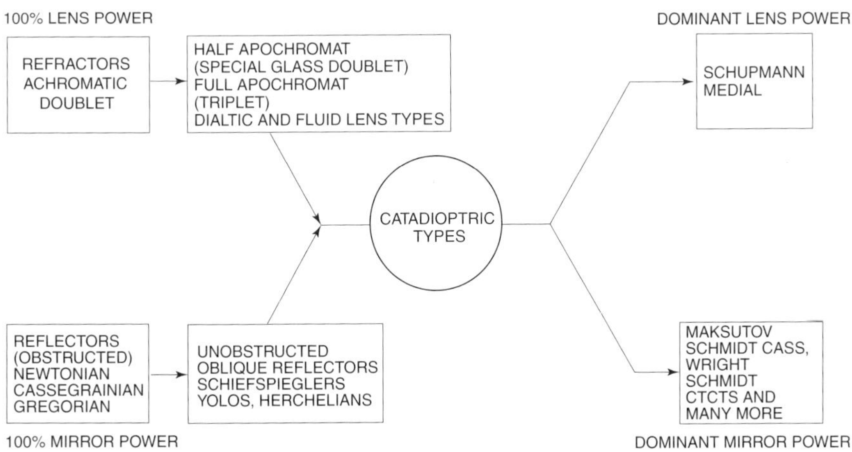

Fig. 3.1 *Telescope design flow diagram.*

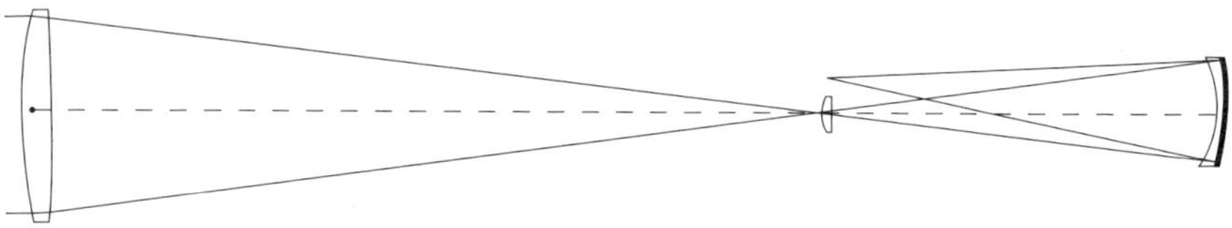

Fig. 3.2 *Optical diagram of the conceptually simplest medial. The thickened line on the back surface of the meniscus corrector represents the reflective coating. The dotted ray path (chief ray) shows how the field element (a lens in this example) images the nodal point of the objective onto the silvered vertex of the corrector, thus optically uniting the main elements.*

Element 1 is the telescope objective and is a single-element positive lens. Element 2 is, in this example, also a single-element positive lens and is located at the mean focus of the objective. Called a field lens, it images the objective onto element 3, the corrector or, if you like, the compensator, which is (ideally) made from the same glass type as the objective, usually BK7. In this example the corrector is a single-element negative meniscus lens. The first thing one may notice is the use of a reflecting coating deposited on the back surface of the slightly tilted corrector lens. Its purpose is to return (reflect) the refracted rays nearly upon themselves, whereupon they are refracted a second time, so that a final image is formed alongside the field element. The image is now easily accessible with a tiny diagonal mirror. We sometimes (inappropriately) call the corrector a Mangin, a term usually reserved for a primary imaging element. Because there are both refracting and reflecting surfaces, Schupmann called his system "The Medial Telescope" (Die Medial-Fernrohre), meaning a telescope design that is between extremes (refractor and reflector) or occurring in the middle. One could also label the medial, as Baker did, a "catadioptric refractor".

To understand the way these elements work together to form a highly color-corrected final image it is best to break down the functions and properties of each in turn.

Examining Figure 3.3A one sees that the axial color of the simple objective is positive, that is, the red focuses long or further from the objective and the violet, being more strongly refracted, focuses short or closer to it. This lens is optically bent to give coma-free images at its focus. In Figure 3.3B we see the corrector used in its usual way as a unit magnification relay (unit magnification relays are inherently coma-free). In this example the corrector is illuminated with a white-light point source. The corrector is tilted to bring the focus to one side. Because the corrector has negative refraction its color sign is opposite that of the objective, that is, the red rays focus short and the blue rays long. By careful adjustment of both the refracting and reflecting surface radii, a corrector can be designed to precisely compensate the axial color of a *particular*, same glass, singlet objective. The lines joining both figures demonstrate the principle. This compensation can be highly precise over a fairly wide range of corrector diameters for any given diameter objective.

Very small correctors require steep refractive surfaces often requiring the in-

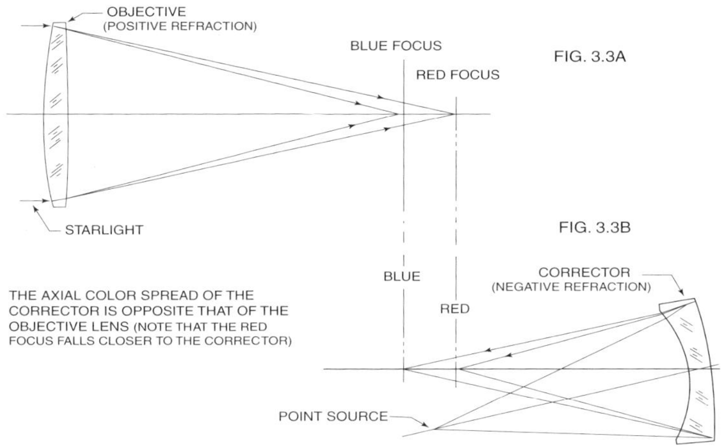

Fig. 3.3

troduction of compound forms. It would appear that all that is needed is to place the elements (objective and corrector) on a common centerline and a color-corrected image will be formed at the location of the point source in the Figure 3.3. Alas, the optical performance is pretty bad when this is tried. First off, beam-walk at the corrector from oblique (off-axis) rays gives rise to strong lateral color. Lateral color is a variation of magnification as a function of wavelength and makes the star images into short spectra, all radiating away from the optical axis. Secondly, since the beam patch (footprint) for each wavelength has a different diameter at the corrector, the refraction angle is not appropriate for the corresponding point on the objective, but for one wavelength. Schupmann showed that the axial color correction in this simple, but flawed, set-up was about ¼ that of the standard crown-flint doublet of the day.

His final brilliant stroke was the addition of a field element—the field lens or field mirror. Now the objective could be precisely and sharply imaged on the corrector for all colors. Lateral color disappears and axial color becomes vanishingly small, in fact 1/20 that of the standard doublet when light flint is employed for the corrector, and 1/100 or better for a same-glass solution. The field element must be designed and positioned rather carefully to achieve the best possible color correction.

Baker reasoned, on theoretical grounds, that the second nodal point of the objective should be imaged on the vertex of the reflecting surface of the corrector. This, in effect, optically unites the two elements in a way not physically possible otherwise. This unique feature of the Schupmann also makes possible an instrument of the highest astrometric precision. (See Footnote 8 on page 10) As an amplification on the above: the second nodal point of the medial's objective, which is shaped for zero coma, lies internally about 40% of the lens thickness from the front surface. It is the point about which a lens may be rotated about without motion of the focused image.

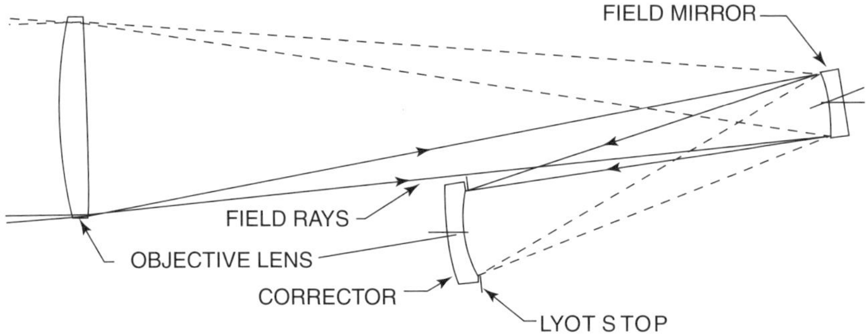

Fig. 3.4 *Imaging the aperture with a field mirror.*

The field element can be a lens or concave mirror as Schupmann pointed out in his 1899 US patent. Whether the telescope is configured with a field lens or field mirror, the imaging performance, neglecting field curvature, is the same, although in some cases the field lens will need to be an achromatized doublet. Whether to use a field lens or mirror is usually decided by the type of research planned, and various schemes are discussed throughout this book. The standard configuration detailed herein uses a field mirror. Figure 3.4 is a simplified sketch showing the action of the field mirror in controlling the field rays. The fine points of nodal point imaging are not indicated and, of course, the image of the aperture stop at the Lyot stop will be slightly, but harmlessly defocused. It helps, in this regard, to position the objective's aperture stop slightly skyward of its first surface. The Lyot stop is principally required for solar prominence studies using an occulting mask (see Chapter 7). It is slightly undersized to block the bright diffraction edge of the aperture stop and, of course, defines the telescope's true (effective) aperture. For further reading, the next chapter details the various optical aberrations that arise and their level of correction in the medial. Next is a brief outline of the medial's favorable *operational* points and some potential practical difficulties. Most are discussed in various levels of detail in later chapters.

3.1.1 The Medial's Strong Points

1. As with any refractor the aperture is high above the observatory floor thus keeping the optical path protected from the worst turbulence, which is near the ground. This is the principal reason large (long) refractors more consistently show diffraction-limited performance.
2. The observer and/or his CCD camera also create turbulence. However, their location at the bottom of the tube, well away from the aperture, minimize thermal degradation of the image.
3. The unobstructed aperture provides the best (highest) possible contrast factor.
4. The medial's optical system is essentially coronagraphic in nature and fully adaptable to solar prominence and other special observations.

5. Atmospheric dispersion (differential refraction) is easily compensated by slightly tilting the field mirror or translating the field lens. These adjustments are handy to the eyepiece location in any medial and are used routinely to keep the instrument in perfect alignment
6. Very low background scatter.
7. Ghost images (flooding) by the objective's 1/6 ghost and the corrector's 1/10 ghost, are spread beyond detection at the eyepiece location.
8. Optics are easily cleaned which helps maintain a high-contrast image
9. The intermediate focus provides a location for special occulting masks. For instance, one can view Jupiter's moons without the mother planet's glare; and because the edge diffraction at the objective's aperture is imaged onto the Lyot stop, the field is dark indeed.

3.1.2 The Medial's Practical Difficulties

1. The optical path length of a typical medial is twice that of an equivalent *f*/number refractor, thus increasing the possibility of tube currents and stratification effects harming the image. Air currents tend to be forced into the beam rims as they travel past baffles. This can be largely overcome by making the baffle openings oversize where possible and also by providing openings in the baffle plate corners away from the beam cones. After a long period open to a *very clear* calm night sky, the tube surface tries to fall well below the surrounding air temperature. This radiational cooling effect is strongest with very low dew points. The author has carefully measured the internal air temperature of his own medial and often finds it 4 to 6 degrees Fahrenheit below that of the outside air under these conditions. This delta T drives a slow convection process that can be seen in the out-of-focus image pattern. Space blanket material, loosely wrapped around the tube, has been shown to mitigate these bothersome currents. Painting the tube aluminum is another easily implemented solution. All this, of course, lowers the tube's emissivity. Air stratification in the tube can be a problem when viewing at low elevations, especially in the early evening before everything reaches equilibrium. Small doors in the tube can flush out the warm air making very early twilight work more effective. Opening the dome or roll-off roof at sunset is also very effective at dissipating trapped daytime heat. All in all, the medial is very well behaved especially when these simple precautions are observed.
2. Touched upon throughout this book is the matter of the thermal settling time of the crown glass corrector. Primarily noticed during polishing and figuring of larger correctors, the effect is *sometimes* noticed in an observational situation. Think back to the days when plate glass was the only material available for the amateur's mirror and how the settling time between spells of figuring slowed progress. The introduction of low-expansion Pyrex helped immensely here. Now, in the case of the Schupmann

corrector, a similar but even stronger potential thermal problem exists because the reflection occurs in a high-index medium making it 1.5 times more sensitive. (Separated mirror correctors are immune to such effects and are discussed in Chapter 8). As with many things of this sort the imagined operational difficulty is far less in the real world. Nevertheless, the medial will, in some situations, show some slight asymmetry in the intra- and extra-focal image for a while during cool-down. Most scopes do!

3.2 Various Optical Layouts

3.2.1 The Medial

Typically the medial's optical path length is appreciably longer than its focal length; for instance at the extreme practical limit, systems utilizing full-aperture correctors will have a light path three times their focal length. Even so, simple optical arrangements can constrain the tube length to just slightly more than the focal length, and in a few folded layouts considerably less. Generally speaking, medials employing a field mirror are rather limited in the arrangement of the optical components; however, the field lens system opens the door to many layout possibilities.

The following sketchpad of optical arrangements typifies those implemented over the years. Although, other practical layouts may be possible, the schemes presented here are sound and will provide the designer and builder with stepping-stones for creating variations in detail to satisfy their instrumental needs. Please note that the sketches are highly exaggerated for clarity and show impossibly fast focal ratios and unworkable tilts! Thickened lines denote a reflective surface. Whenever possible, a realized layout will give credit to the designer and, as with untried schemes, give the rationale for the choice.

We begin with the Standard Design shown in Figure 3.5. This design is constrained by the requirement to minimize the tilt of the field mirror and corrector respectively. The field mirror tilt (nominally 3 degrees) is minimized to a limit case where the special baffle is just workable. Although the impact on field center aberration control is somewhat insensitive to field mirror tilt, ray trace studies show that unnecessary tilt should be avoided. Corrector tilt, as with any configuration other than the split-field scheme, is needed to make the focused beam accessible. For research, this tilt should only be large enough to satisfy the scientifically useful un-vignetted field based on ray trace studies of various objective-to-corrector diameters and *f* ratios. Most successful general purpose Schupmanns never exceed corrector tilts of 0.75 degrees, usually much less. This is extensively discussed in later chapters. The drawing shows (dotted in) the practical limits for corrector diameters either side of the all-spherical Super Schupmann. The advantages and drawbacks of varying this diameter parameter are also covered in later chapters. Usually a diagonal mirror is interposed to pick off the beam

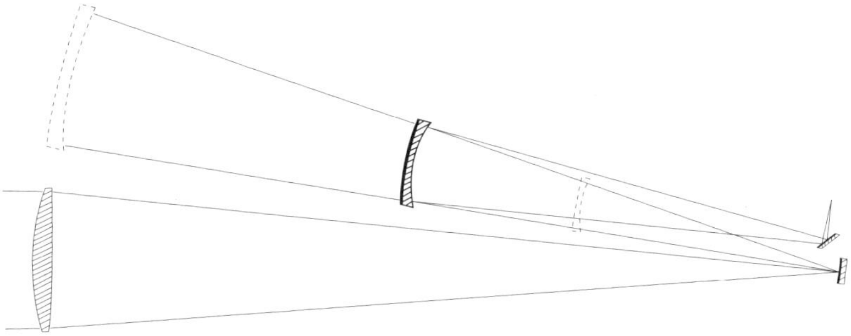

Fig. 3.5 *Schematic layout of Standard Design Schupmann.*

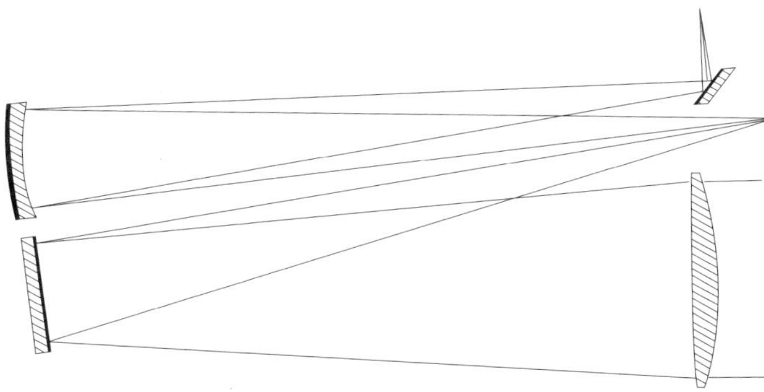

Fig. 3.6 *Logan's folded Super Schupmann.*

just short of focus. This makes for a very enjoyable viewing situation in the Standard Design as the observer is at the back of the tube away from the objective as with any refractor.

One variation of the Standard Design layout is Gerry Logan's folded version shown in Figure 3.6. In this version the observer views as with any short-focus Newtonian, usually with no stepladder needed. A photo and description of this instrument are found in Section 12.6.

Moving on to the field lens systems we see many layout possibilities. Figure 3.7 shows the in-line system. To the author's knowledge, only three instruments of this layout have been implemented. Two employed tilted components and the third was an un-tilted split-field type. Figure 6.4 shows the region of the field lens for the split-field. The in-line idea is limited to smaller correctors to keep the overall tube length reasonable. The author's 7.5-inch *f*/13.3 solar/planetary medial, seen in Figure 12.3, is about the practical length limit.

Figure 3.8 shows Anton Kutter's arrangement for his 5.9-inch *f*/11.3 planetary instrument, the "Else Kutter Medial" seen in Figure 12.4. The drawing is the author's tracing of a sketch from a private communication with Kutter. The tube is outlined in this sketch to show the skewed path, which makes good use of internal space for much of the corrector's light path. Although never previously imple-

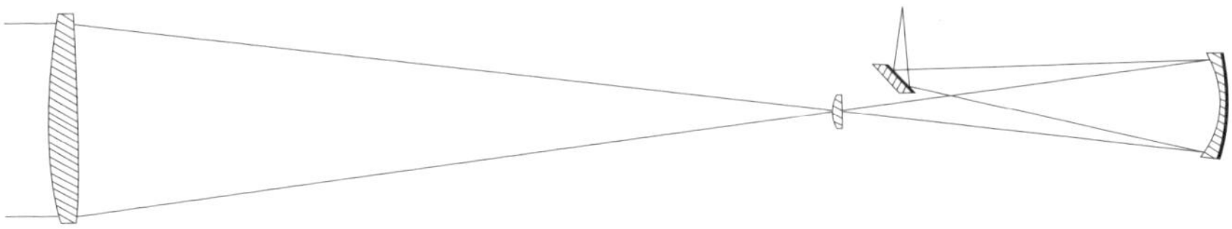

Fig. 3.7 *Basic "In-Line" field lens layout. Limited to smaller or specialized solar medials.*

mented, Schupmann favored this skewed objective path for his elbow setup (see Figure 1.5, drawing numbered Fig. 6). This layout is a basically a fine idea, however, the builder should be aware that the skewed path of the converging cone from the objective places one edge of the cone in a more or less permanent boundary layer of air clinging to the tube wall. Unless adequate clearance is provided, an asymmetry of image structure may be noticed due to "tube seeing" effects. It is hoped that other amateurs will give this layout serious thought. The photo insets in Figure 3.8 show the neat, compact, back plate mounting of the field lens, flat mirror and focuser of Kutter's medial. Kutter's beautiful beam-aiming "spider" is made of flex straps that are deflected with two push screws, thus centering the pupil precisely on the corrector. By studying Figure 3.8 (Lower Right) you can see how this motion is realized. You will also notice how the medial does not lend itself to standard commercial components without some cutting-away or other modification. The need for very close beam spacing in the focal region almost always requires inventive design and good machining—nothing to concern the advanced ATM, however.

Schupmann's classical "elbow" arrangement (not shown here) can be seen in both the German and American patent drawings (see Figures 1.4 and 1.5). Limited to small (two-element) correctors on instruments with very small fields, this form has been largely abandoned. The last two instruments constructed using Schupmann's elbow idea were Kutter's 122mm *f*/10 "Berta Kutter Medial" (1936) and his 270mm *f*/10 instrument. Today only one operational "elbow" system exists, the 323mm *f*/11.5 Stuttgart medial (see *Sky and Telescope*, March 1983).

Figure 3.9 illustrates James Baker's design for replacing Harvard College Observatory's 15-inch great refractor with a true apochromat. This very unusual approach for "getting around the bend"; so to speak, was employed to provide clearance at the original tailpiece by enabling the corrector tube to lie alongside the old tube. As detailed in Chapter 1 the instrument was instead mounted at Harvard's Sacramento Peak facility. The major advantage of the idea is the ease of baffling the light path, as the long BSC2 glass "block" completely isolates the corrector in its own separate tube. Optically it simulates the in-line and elbow field lens configurations. Baker achromatized the field lens, which allows a higher degree of color control than a singlet, especially when located well beyond the hydrogen-alpha focus as in this design. The cemented doublet is, in turn, cemented to the prism. (A singlet field lens gives rise to secondary lateral color; almost in-

Fig. 3.8 Top *Anton Kutter's optical arrangement for his "Else Kutter Medial"*. Lower Left *Kutter's tailpiece plate showing flex "spider" push screws.* Lower Right *A view down the focuser axis as seen from the corrector. Note the closeness of the elliptical fold mirror and eyepiece centerlines. The singlet field lens is in the uppermost cell.*

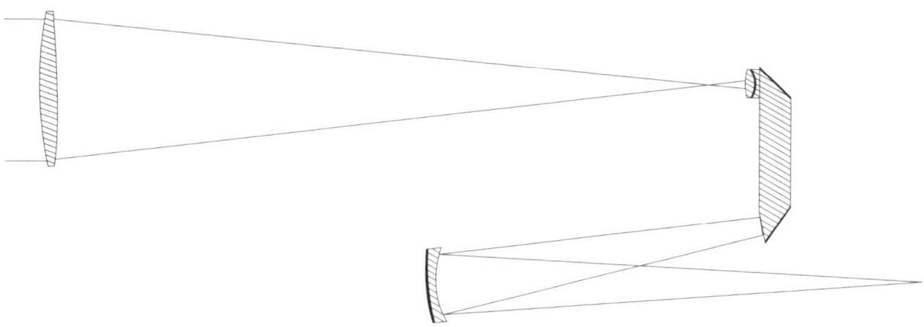

Fig. 3.9 *The optical layout of the Sacramento Peak Solar Schupmann.*

significant for most designs.) The design holds up fairly well in ray trace analysis and there is no problem adjusting for the prism's long, 10.93-inch, in-glass path. Intended to be used with an occulter, the large prism is shielded from the heat and light of the Sun. Because the instrument apparently had optical problems not associated with the design, the layout is a serious contender for large aperture, field-lens systems. The author would love to see these "stored" optics brought to life in a field-mirror configuration for solar studies.

Figure 3.10 shows the design layout for the author's 14.1-inch *f*/12.8 medial. Here again we can use the term folded in its true sense in that the tube length is

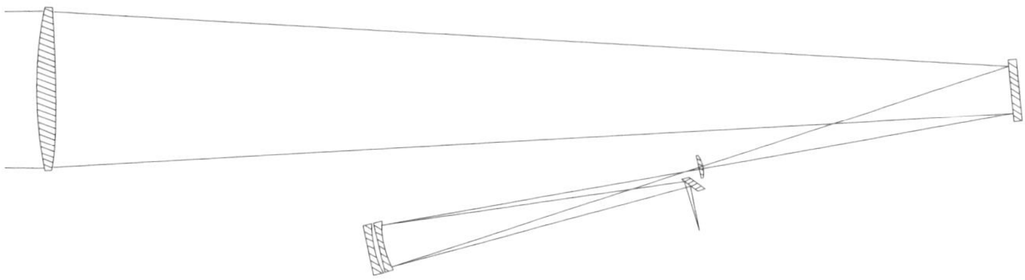

Fig. 3.10 *The author's design arrangement for a 14.1-inch f/12.8 folded medial for double star research. Note the separated reflector of the corrector element. (See Chapter 8.)*

Fig. 3.11 *The author is seen holding his finished 14.1-inch f/12.8 BK7 objective and cell next to the objective of his current 9-inch f/11.1 instrument. In the reflector world this would not be considered "aperture fever," however, in amateur refractor circles it represents a very bad case of the disease. The need for higher resolution did not drive the aperture choice as the 9-inch is perfect from that standpoint given the typical New England seeing and the authors double star astrometry program. Instead, it is in the work of differential (Δm) photometry of faint pairs (12–13 mag) that drives the aperture up. This is because the stars are greatly dimmed by the standard photometric BVRI filters employed in this work, thus starving the CCD of light. Keeping the exposures reasonable in these color bands substantially reduces the effects of mount and seeing wander. Although, the light gain is just under one stellar magnitude, it is well worth it. Naturally, on the very best (rare) nights, doubles with separations under 0.6 seconds of arc will be accessible for measurement. Keep in mind that visually resolving (splitting) a close pair is one thing; however, measuring the position angle and separation to a high accuracy with a CCD is quite another. The new instrument is scheduled to be operational in late 2007.*

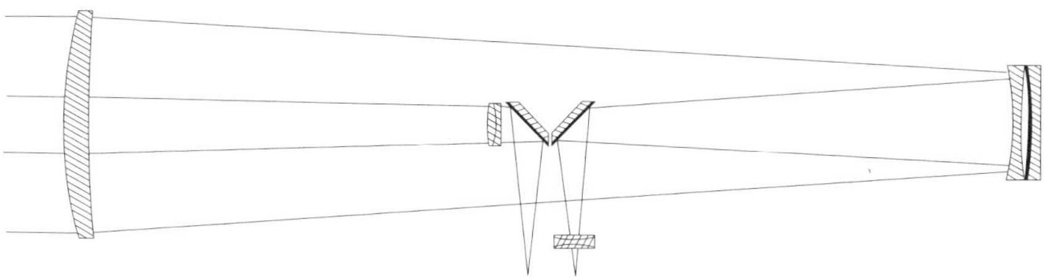

Fig. 3.12 *Optical layout of the Rathenow "Brachymedial." This instrument is extensively discussed in Section 9.9.*

shorter than the focal length (neglecting, of course, Dialyte type systems). This idea places the final focus conveniently off the observatory floor while the tube's bottom end just clears it. The fold mirror flat is made from Cer-Vit, a no longer produced, zero expansion, vitreous ceramic. By folding the light path in this manner, the instrument will fit in a 10 x 12-foot roll-off-roof observatory.

3.2.2 The Brachymedial

Although not a medial in the accepted sense where the corrector acts as a unit magnification relay, the brachymedial, or shortened refractor, idea offers many interesting possibilities. In general the tube length is about half the focal length of the traditional form. The early versions of this optically obscured design were usually intended to work at about *f*/20. Limited to small variations in the final focus position and the makeup and diameter of the corrector itself, these old designs were pretty much look-alikes. The patent drawings shown in Chapter 1 give some idea of the basic system. The sketch shown in Figure 3.12 depicts the Rathenow instrument which is the only presently operating brachy. The reader will find detailed information on this instrument and other old designs, as well as the latest brachymedial developments in Chapter 9. The drawing is highly foreshortened to show more clearly the element details, and is loosely based on designer/builder Edwin Rolf's original sketch from his 1953 technical report, "Der Verkürzten Refracktor" for the Akademie der Wissenschaften.[1]

3.3 Schupmann's Dream

The possibility of color correcting large, single-element objectives led Schupmann to apply his design ideas to instruments in the range of 1.5 meters (~60-inches). The following detailed drawings published in Schupmann's *Die Medial Fernrohre* give us a good idea of his thoughts. Unlike modern design interpretations, Schupmann worked to minimize the corrector diameter. Although, carried to extremes, it is a perfectly understandable goal for the apertures he was considering. Generally he preferred correctors about ⅛ the objective's diameter, thus an aperture of 60-inches leads to a corrector of only 7.5-inches—none too big for an amateurs 10-inch medial! As discussed elsewhere, Schupmann achieved good de-

[1] W.P. Hartmann, Private communication, 1993.

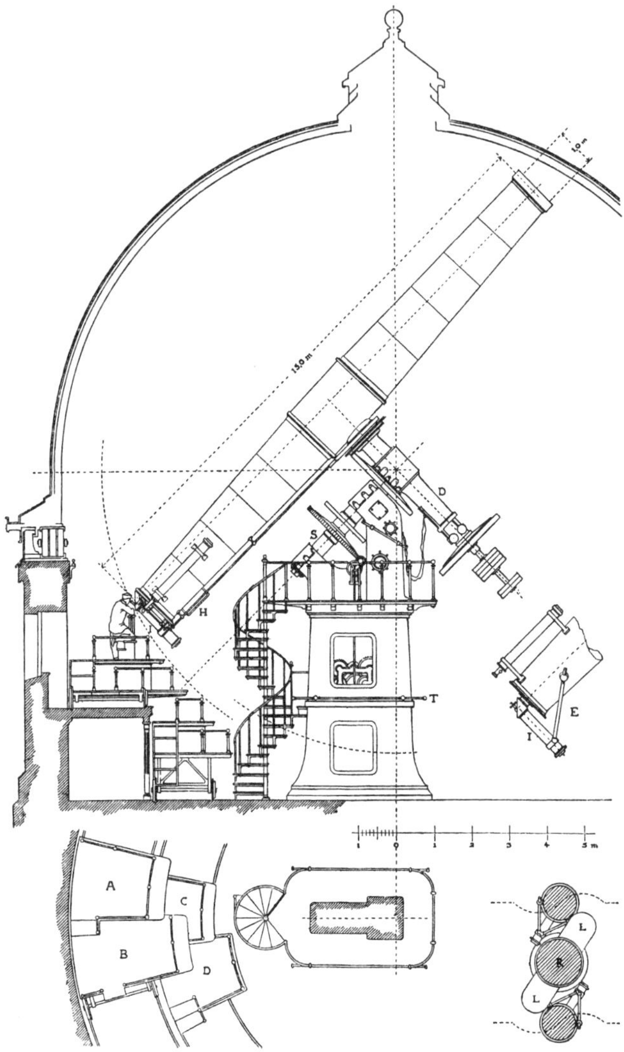

Fig. 3.13 *This drawing by Schupmann shows his overall design for a 1-meter f/15 medial.*

sign results by employing multi-element correctors of light flint.

When Schupmann did this design visual observations were still the dominant method of collecting data, and we see that in these drawings. Even through photography was rapidly improving, for lunar, planetary and double star studies the eye was still the supreme detector. Large reflectors of the era required frequent resilvering giving Schupmann further impetus to develop a huge, perfect refractor (the silver coating on Schupmanns Mangin-like last element was protected from tarnish).

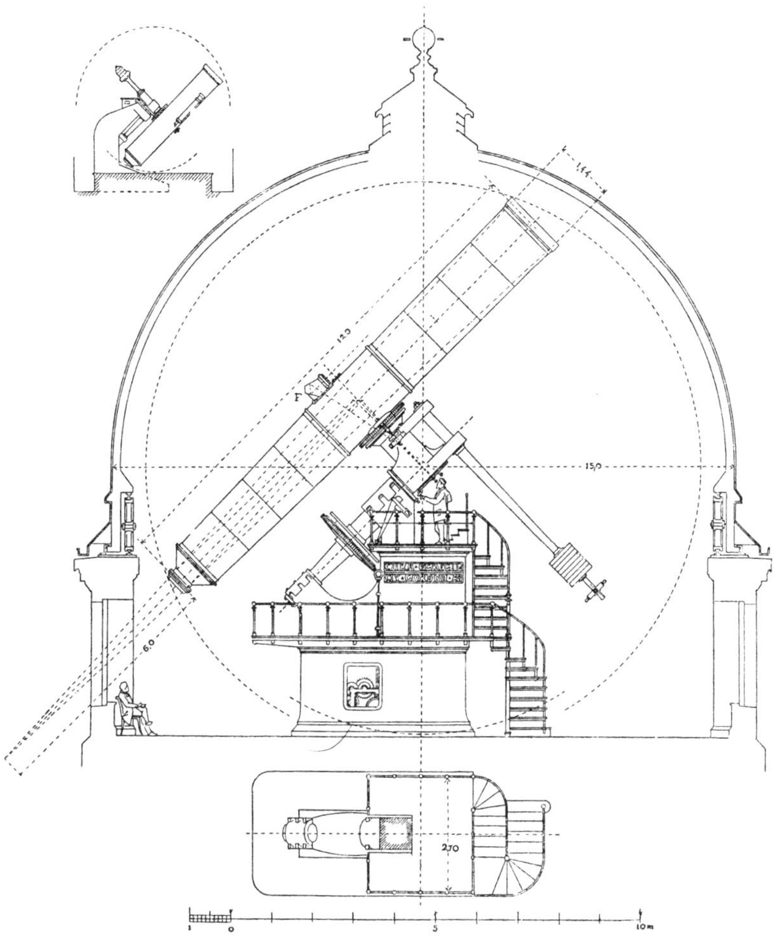

Fig. 3.14 *Schupmann's design drawing for a 1.44-meter brachymedial and observatory. This was to be the largest refractor in the world. Interestingly, a close (half scale) model was actually completed and is described in Chapter 9. Note the offset counterweight shaft allowing a semi-fixed eyepiece location at the end of the short, hollow declination axis. The finder combines an achromatic doublet and a "penta" mirror arrangement to send the wide-field view cross-axis to focus beside the high power eyepiece location. Contrast this finder set-up with the Rathenow Brachymedial layout in Figure 3.12.*

Figure 3.15 shows an overall view of his proposed tailpiece. His clever use of a simple flex-plate in the 90 degree push screw alignment mechanism is clearly seen. The baffled adjustable tube pivots well up into the inside of the main tube.

Finally, in Figure 3.16 we see the fine detail of his more developed tailpiece design including the two-element corrector. Note the 90-degree prism with integral field lens.

This design layout gives us Schupmann's so-called elbow arrangement, a feature that was employed in all the early medials until Anton Kutter broke the

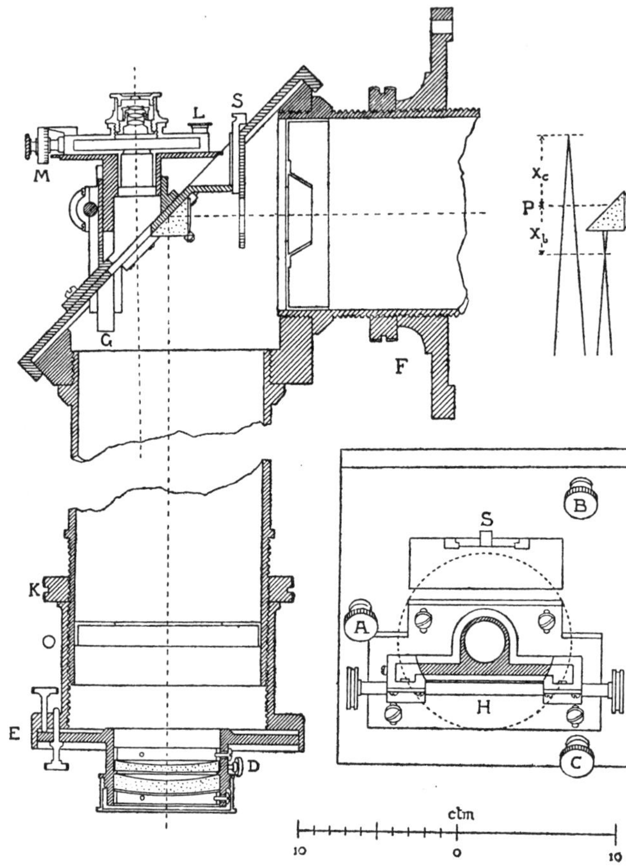

Fig. 3.15 *This illustration by Schupmann shows the proposed tailpiece for a 1-meter f/15 medial.*

trend (see Figure 3.8 on page 39). Also note the close separation between the beam leaving the prism and the returning beam entering the eyepiece tube. This is required to minimize the corrector tilt. Excessive corrector tilt would spoil the instruments use in filar micrometer measurements of binary stars as field distortion would be difficult to calibrate out. (Note the filar micrometer in the lower right hand portion of Figure 3.15.)

Field distortion is not a problem in modern designs as long as reasonable care is taken in designing the structure of the focal region to keep the tilt within bounds as described in this book.

The only remaining operating medial built to the elbow plan is the Stuttgart 12-inch. Modern workers may wish to explore the possibilities of tiny correctors using Schupmann's elbow layout.

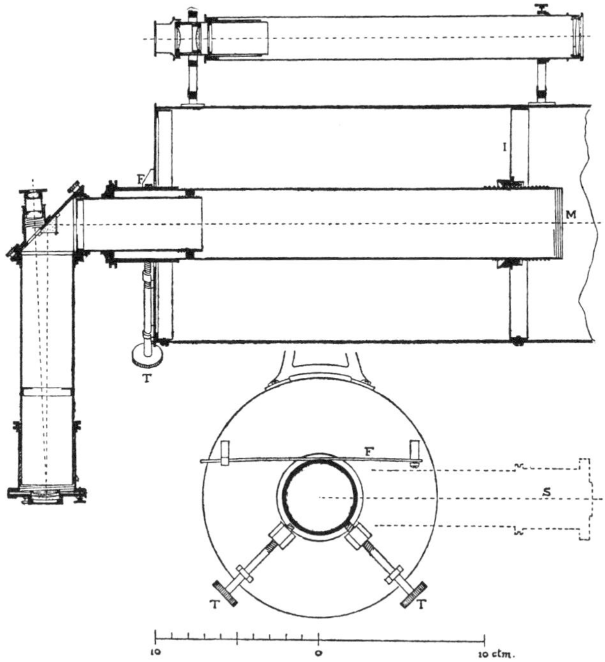

Fig. 3.16

3.4 Schupmann Offshoots

The author would be remiss if one of the real gems to emerge from the Schupmann principle were to go unmentioned, namely the "Baker Refractor Corrector." Here a standard achromatic doublet objective is corrected for secondary color with a corrector made from the same, or nearly the same, crown and flint glass used in its objective. It is employed in the usual way seen throughout this book. James G. Baker proposed the idea at an A.A.S meeting at Haverford College in 1950. Quoting Baker:

> The basic idea of this corrector stems from the Schupmann medial telescope. The author's particular contribution lies in the recognition that the principle of the Schupmann Mangin can be extended in doublet form to correct the secondary spectrum of even the largest refractors.

Amazingly, Baker's powerful concept fell on deaf ears, and no implementation of the idea is known to exist. This is, of course, a sad situation. Imagine the Yerkes 40-inch without a trace of color! The concentration of energy in the image would give the instrument an identical limiting magnitude as a reflector of the same size. Almost every type of research would benefit. Some field-of-view would be given up to hold aberrations under control for astrometric work—a small

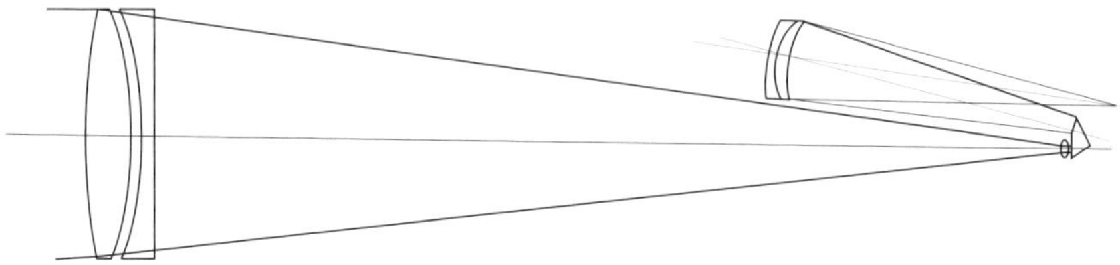

Fig. 3.17

price to pay given the tiny fields needed for most scientific measures. Figure 3.17 shows Baker's optical layout from a prepublication manuscript of his article "The Catadioptric Refractor" (see *Sky and Telescope*, Feb. 1951).

Note that the field lens is cemented to a modified Porro prism—a neat idea even today. Baker mentions that a field mirror may be used instead.

Although no prescription is given in this volume, an advanced amateur will have little difficulty working up a design to correct his standard (Fraunhofer) achromatic refractor. The idea is practical for larger—i.e., 10-inch plus—instruments.

Chapter 4
Optical Aberrations of the Medial

4.1 A Practical Discussion

Schupmann's medial is inherently a narrow-field instrument. We can define the useful field of view in various ways. Obviously, low power visual observation of star clusters makes the least demands on the medial's optical system since the aberrations of the eyepiece and the eye overwhelm those of the telescope, thus preventing direct observation of the detailed nature of the individual star images. For this sort of pleasure viewing we can consider fields of view of about ½ degree as the practical limit, keeping in mind that a scientifically useful amateur's medial will have a focal length of between 80 and 130 inches. We can, for most critical work, define the scientific field of view as coinciding with the diffraction-limited field. The diffraction-limited field is that field in which star images do not sensibly depart from the classic Airy pattern of energy distribution. One of the best ways to define this performance is to examine what is called the "modulation transfer function" (MTF) for various field angles. The MTF is a measure of an optical system's response to 100% brightness modulation of spatially varying targets in object space. This response is nongeometric, but instead is based on the wave nature of light. The OSLO ray trace program outputs the data in the form of the optical system's modulation response versus cycles (lines) per mm in image space. The MTF output plot includes the curve of a theoretically perfect system to help the designer judge performance. Usually these plots display the MTF for three field points: the center, 70% of the field diameter and the field edge. When this function for a stipulated (required) field of view agrees with the theoretically "perfect" or ideal system our telescope can be no better. Small departures from the ideal can be tolerated for most scientific work. The optical designer's task is to minimize aberrations across the field with particular emphasis on developing the best performance in the region around the field center, or so-called planetary field. In later chapters the MTF parameter is used to evaluate the performance and field limitations of various system layouts. However, we must first describe the medial's aberrations that ultimately limit sharp imaging both on-axis and across the field. Some of what is described below is necessarily repeated or touched upon in other sections of this book.

4.2 The Aberrations — General

Briefly, the medial's on-axis aberrations can be made very tiny. Over the wavelength range 0.36 to 1.10 microns a well-designed medial will surpass the best apochromat by a factor of five or more at 1/10th the cost. In describing refracting telescopes we need to examine many aberrations in terms of how they vary with wavelength, because we are dealing with dispersive elements and thus we may encounter such things as variation of spherical aberration with color, or simply "spherochromatism." Naturally, the host of aberrations encountered in refracting telescopes such as longitudinal color, lateral color, spherical, astigmatism, coma and field curvature are also potential hazards in the medial and, where possible, must be eliminated or at least minimized below sensible detection over a usable field.

4.2.1 Spherical Aberration

The Standard Design, which employs a singlet corrector, is free of 1st order spherical aberration with all spherical surfaces. This is a constrained design as both spherical and coma are corrected at only one ratio of objective and corrector diameter. We can bend the objective and obtain a spherical solution for different corrector diameters, however coma will reappear. Designs, discussed later, employing correctors with a separated mirror element, can also be developed as all-spherical systems for other objective-to-corrector diameter ratios. Other systems lying away from the standard design employing a single-element corrector of the same glass as the objective, will require aspheric work which varies from just detectable to twenty waves or more for the tiniest correctors. A graph showing the number of waves of aspheric correction versus relative corrector diameter is shown in Figure 4.1. Interestingly, a specific solution for spherical correction exists for systems employing a crown objective where light flint is used for the small, singlet corrector. Anton Kutter used this design approach for his medials. For his modest apertures, used for visual work only, the residual longitudinal color (C-F) is 1/10 that of an equivalent achromatic objective and furthermore spherochromatism, which normally plagues medials with small (crown) singlet correctors, is totally absent.

4.2.2 Longitudinal Color

The defining characteristic of a well-designed medial is its absolute achromatism. The color correction is visually indistinguishable from a reflecting telescope. A refracting telescope for visual use is said to be fully color corrected when all the colors in the visual spectral region focus geometrically to a spot substantially smaller than the Airy disc. This realistic spectral range extends from 0.43 through 0.65 microns. The standard 6-inch *f*/15 achromatic objective fails this test terribly, and the best examples of apochromatic instruments nearly fulfill this requirement, but fail at the violet end. As an example to show the extremes one must go to with a standard refractor, a 3-inch *f*/56 Fraunhofer doublet just enters the realm of super fine performance across the visible spectrum! One would certainly want to fold that

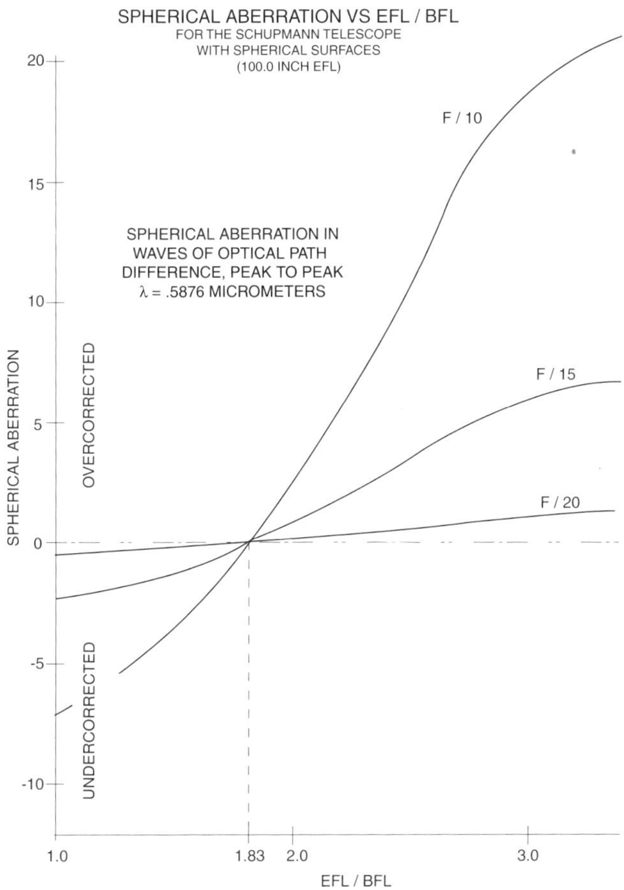

Fig. 4.1 *The influence on spherical aberration for various objective-to-corrector diameter ratios when employing a singlet corrector. Various f/ratios are shown to give the builder an idea of the aspheric work required as one moves away from the all-spherical solution.*

14-foot light path, but what a nice white-light solar instrument it would make when equipped with a Herschel wedge and eyepiece filter!

With the introduction of the CCD we need to extend the color correction to cover the region 0.36 through 1.10 microns. The medial can be designed to cover this range over useful fields of view. The best axial color correction is obtained with systems that employ full-aperture correctors. An 8-inch *f*/15 system can be designed to show axial color reduced to better than 0.02 mm (0.0008 inch) over a spectral range from 0.30 to 1.10 microns. As we reduce the corrector diameter, spherochromatism makes its appearance and at a corrector diameter of about ⅓ aperture we find our diffraction-limited spectral range is limited to the visual region only. Only by separating the corrector's reflecting surface, thus reducing its curvature or lens power, can we find solutions for corrector diameters as small as ¼ that of the objective (as long as the *f*/ratio is *f*/12 or greater).

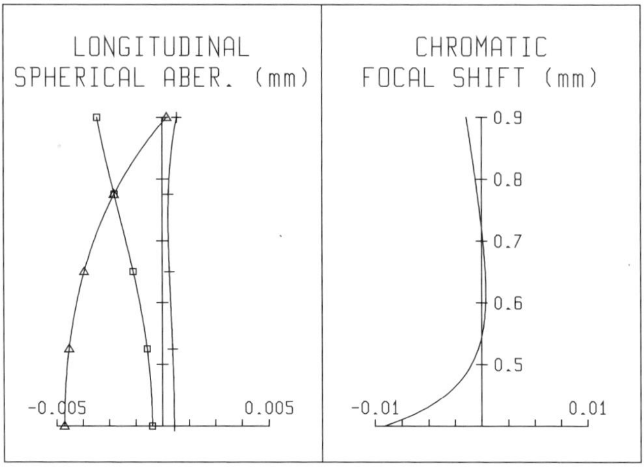

Fig. 4.2 *Spherochromatism and color curve for an 8-inch f/11.25 medial employing a full-size corrector (note scale when comparing other data below).*

4.2.3 Spherochromatism

As touched upon earlier, this aberration imposes serious limitations on small (singlet) corrector systems when a large wavelength span is required. It is simply a variation of spherical aberration with color. The strength of this aberration varies as the square of the *f*/number and linearly with objective-to-corrector diameter ratio, going to nearly zero in a special case described below. Typically, after correcting the system in, say green light, we find spherical overcorrected for blue and undercorrected for red light. With singlet correctors more than ½ the objective diameter, the spherochromatism no longer competes with the Airy disk (at *f*/12 and higher) over a spectral range of 0.35–0.9 microns (typical of an amateur's CCD). At *f*/15 we find correctors as small as ⅓ diameter perfectly satisfactory for planetary fields. For visual use this aberration is far less bothersome since the spectral response curve of the eye is very narrow compared to a CCD. Interestingly, as mentioned earlier, spherochromatism goes to almost zero when a full-size corrector is employed. See Figure 4.2, which shows the residual spherochromatism and color curve for an 8-inch aperture 90-inch focal length example of a full-size corrector system. The wavelengths used for the curves throughout the following examples are: 852 nm (squares), 546 nm (crosses) and 430 nm (triangles).

These "full size" systems are quite practical up to about a 6- or 7-inch aperture, beyond which the corrector becomes more difficult to figure, and thermal settling problems in use can sure take a bite out of the early night's work. All-quartz medials have been made to solve the thermal problem, but are limited to professional systems due to the high cost of *optical grade* quartz. The amateur could consider a modest 6-inch quartz system and, to that end, the author presents a really nice *f*/15.2 design employing a full-size corrector in Section 5.6.

As touched upon earlier, spherochromatism can be reduced by a factor of

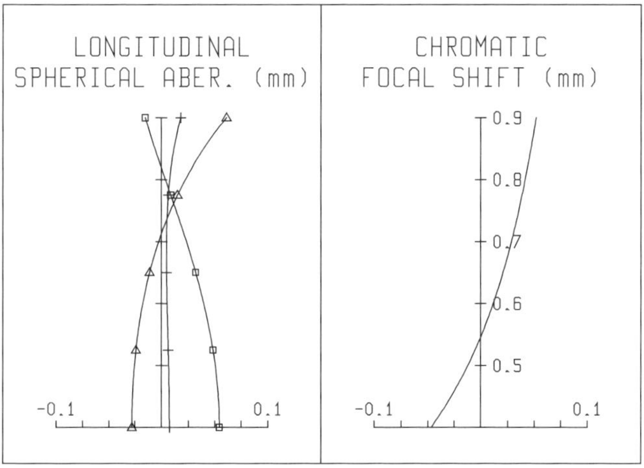

Fig. 4.3 *Spherochromatism and color curve for an 8-inch Standard Design working at f/12.5 (note scale).*

two by separating the reflecting element in the corrector, thus putting both surfaces of the lens in double pass. This stratagem gives the designer one additional degree of freedom. Now, by double passing, the individual lens powers are reduced and, by optically bending the lens, a solution for spherical can be found for nearly any corrector. Unhappily spherochromatism, although weak, is not necessarily minimized simultaneously so the designer may want to aspherize in favor of reducing spherochromatism to its weakest value. Field curvature remains the same as in the singlet case; however, chromatic variation of astigmatism disappears when adapting the idea to a 54% diameter corrector, thus giving a *geometrically* sharper "Best Focus." There are substantial practical advantages—for instance, the reflecting element can be made of a zero expansion glass greatly improving the focal length stability. Also the reflection is now in air and much less sensitive to small-scale defects. By placing the mirror element in rim-to-rim contact with the lens its coating is protected, especially if thermostat control is employed during periods between uses. Obviously, each such system should be individually examined for the strength of ghost images to keep them below visibility. In particular, watch the second surface of the corrector to be sure that it is not too close in radius to the mirror, or a weak, highly defocused ghost may appear! Figure 4.3 shows the residual spherochromatism and color curve for the Standard Design of 8-inch aperture and 100-inch focal length (*f*/12.5), and Figure 4.4 an otherwise identical instrument employing a separated reflecting element in the corrector. Crossing points were adjusted for the 70% zone of the objective for best performance. The MTF curve (not shown here) for both arrangements is identical and theoretically perfect. A design prescription for a separated-mirror type corrector is given in Table 8.1.

Finally, we have an example employing a BK7 singlet corrector only 34% of the objective diameter. The *f*-number and aperture and wavelengths traced are

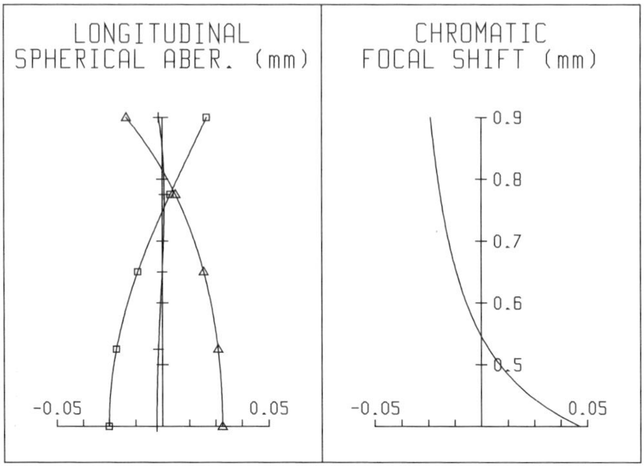

Fig. 4.4 *Spherochromatism and color curve for 8-inch f/12.5 Standard Design modified to employ a separated reflecting element in the corrector (again note scales).*

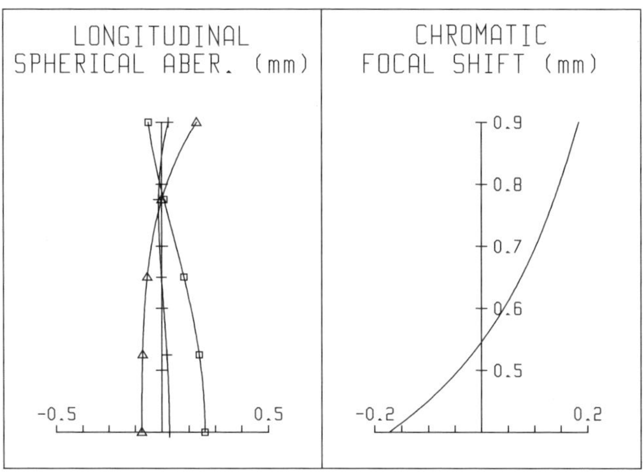

Fig. 4.5 *Spherochromatism and color curve for an 8-inch f/12.5 medial employing a 34% diameter corrector.*

the same as in the last two examples. The first surface of the objective was aspherized to correct spherical aberration (see Figure 4.5). This is a limiting case, because the spherochromatism now begins to noticeably degrade the MTF. Increasing the aperture only 1-inch (holding the focal length constant) corrupts the performance totally; however, at *f*/15 (6.6-inch aperture) the performance is, MTF-wise, perfect.

As mentioned earlier, one can practically eliminate spherochromatism and also obtain an all-spherical solution for the case of a small singlet corrector by employing a light flint in combination with the normal crown objective. This leads to a corrector approximately 27% of the objective diameter. Longitudinal color, although quite small, limits the design to visual use. Also because of the impossibly large working angles for the tiny corrector the field is forced to be necessarily very small. Thus this is, in general, an impractical solution especially when considering wide-band CCD observations. All in all, the best approach to attaining high performance with tiny (¼ diameter) correctors is to utilize BK7, as in the objective, separate the reflecting function and work at higher *f*/numbers. This scheme even allows an all-spherical solution for ~40% diameter correctors.

4.2.4 Lateral Color

Lateral color manifests itself as short, radially disposed, spectra for star images located off axis, The further off-axis, the greater the length of the spectrum. It is, in reality, a variation of magnification with color. Astrometric work over a wide spectral range would be severely compromised if this aberration were present in any but the tiniest amount. This aberration is fully corrected in the medial by Schupmann's brilliant introduction of the field element, which optically superimposes the objective on the corrector, thus reducing oblique-ray "beam walk" to zero. This design feature can be so precisely implemented that lateral color is reduced to less than 0.0002 mm at the edge of a ½-degree field! This field element is kept perfectly aligned by an adjustment near the eyepiece. One can also purposely introduce non-radial "lateral color" uniformly across the whole field, thus providing one of the most unique capabilities of the design, namely, the compensation of contrast-robbing atmospheric dispersion for low lying objects. This is accomplished by slightly tilting the field mirror (or translating the field lens) in true elevation space. This adjustment de-centers the optics, thus introducing "wedge" or "prism." The net effect is to cause the instrument to behave as if a weak objective prism were in place, rotated to oppose the atmospheric dispersion direction. In practice one simply tweaks the mirror tilt until the short star spectrum "disappears". As the apposing spectrum direction and strength is a function of RA in an equatorially mounted instrument, the adjustment may have to be repeated for a long observing session on a given low-lying object. Although this dispersion compensation is the same for all field points, the field is so small that the difference in atmospheric dispersion across the field (at any given zenith distance) is harmless. BK7 and air have different partial dispersions, but the atmospheric dispersion compensation—though not perfect—is so nearly correct that the residual is practically undetectable.

4.2.5 Coma

As with any refractor, coma is an aberration that can be fully corrected. The medial employs a color-correcting relay element, which works at 1-to-1 or unit magnification and is by itself coma-free. The single-element double-convex objective lens can also be made coma-free by shaping or "bending" the lens appropriately. For BK7 glass a front-to-back radii ratio of around 10 with the front surface the more convex reduces coma to near zero. Overall field performance can, in some layouts, be somewhat improved by deviating slightly from a strictly coma-free objective.

4.2.6 Astigmatism

Astigmatism is a dominant aberration in the standard refractor as well as most tilted component telescopes and the medial is no exception. The aberration fundamentally limits the diffraction-limited field of view. In the standard medial we can fully correct this aberration for one point in the field, usually the field center. In the case of the untilted version or "split-field" design, the astigmatism is radially symmetrical about the parent axis. However, in the standard design, which employs tilted elements, astigmatism is non-uniform about the field center, as no parent axis exists. Astigmatism increases in strength as the square of the field angle, making it one of the most corrupting of optical aberrations. Full-aperture correctors require the least tilt to access the image and they work over the smallest field angle for a given true field, and for that reason exhibit the least field-edge astigmatism. Despite the horrors of this aberration, diffraction-limited fields of ¼ degree or more are achievable with the standard medial operating in the region of *f*/12. Bert Willard has designed an astigmatism-free medial that lends itself to the un-tilted split-field design. This anastigmatic medial is described in Section 6.3.

4.2.7 Curvature of Field

The medial has fairly strong field or Petzval curvature, which is negative or backward bending; *i.e.,* concave to the eyepiece. The curvature varies linearly with corrector diameter for a given objective, with full-aperture correctors showing the shallowest curvature. Systems that employ a field lens will show less curvature compared to a field mirror layout. This is because the field lens is thicker in the center and acts as a partial field flattener, whereas the field mirror is concave, thus aggravating the curvature. The objective alone has a very shallow field curvature and slightly cancels that of the corrector. The field curvature of any Schupmann *objective* is equal to its focal length times its refractive index and is concave to the objective lens. Thus the field curvature comes mainly from the corrector—specifically, its reflecting surface, and is roughly equal to ⅓ of its radius for a field mirror system and one half its radius for a field lens system. For a 7-inch aperture *f*/12 field-mirror instrument this curvature radius falls in the range of 400mm to as steep as 63mm for a 20% diameter corrector. For the standard all-spherical design of 100-inch focal length the field curvature radius is 224mm or 8.818-inches. Given this, we always find that the best overall field performance (after correcting for field tilt) is achieved when the system is focused back slightly such that the sagittal and tangential astigmatic fan is, on average, equally split all around the field edge. The geometric on-axis image grows slightly but remains diffraction-limited at this compromise focus. This is all easy to do in a ray trace program, however, under the stars in the presence of turbulence etc., we simply focus the best we can and get on with the task of data taking. When the Standard Medial is used visually it is virtually impossible to detect this curvature especially if the eyepiece axis is adjusted to compensate for the tiny field tilt. Adding a field flattener lens very close to focus will work, but the astrometric qualities will be compromised as the dis-

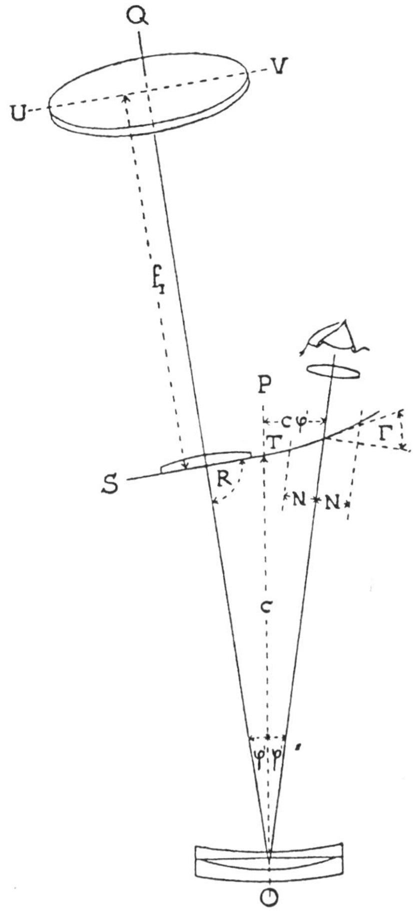

Fig. 4.6 *Schupmann's drawing showing field tilt and curvature. From "Die Medial-Fernrohre."*

tortion rises to ~1% at the edge of a ½-degree field. Implementing such a lens is, in actuality, impossible for CCD work, as it needs to be within the windowed chamber and nearly in contact with the chip! In general, the limited size of consumer CCDs makes field flattening unnecessary.

4.2.8 Field Tilt

This "aberration" is easily compensated by tilting the focal plane of the CCD or eyepiece. The tilt in the Standard Design medial is very small (on the order of 1.4 degrees) and can often be ignored. The author's 9-inch medial employing a ½ size corrector had a measured field tilt of 1.5 degrees before shimming it out. The wonderful, although necessarily exaggerated, drawing by Schupmann in Figure 4.6, clearly shows the curvature and tilt of the field. It is interesting that he used a simple field lens in the example. Schupmann was well aware that the field curvature was "curled up" more at the far edge, thus not a true spherical surface. Note the orthogonal tilt of the objective lens to correct astigmatism. Anton Kutter used very small correctors in his medials and the field tilt and curvature were quite strong, although harmless, viewing planets. He says of this:

> I get a power of 85 times and therefore a true field of 28'. On this relatively modest field I can clearly perceive the toroidally deformed field.

In modern times we steer away from such configurations. Also, by tilting the eyepiece we can keep, as explained next, the effects of tilt and curvature well within the depth of focus.

The best way to detect field tilt is to knife-edge null the system on a fairly bright star in the following way: center the star and remove the eyepiece, then adjust the focuser such that the aperture grays evenly at the field center when resting the knife-edge on the drawtube face. Scribe the drawtube edge. Move the telescope in declination to bring the image to the upper field edge that you have determined as your working scientific field. Now, using the focus knob, null and scribe again. Be aware that you are nulling an image with slight astigmatism and that it is advisable to cut the beam from different directions and average the results. Repeat for 3 additional 90-degree disposed positions. The difference in scribe mark locations will indicate the focal plane tilt so the builder can develop a shim schedule. For systems employing a full-size corrector the major tilt is in the plane of the corrector tilt. This is defined as X. Tilt in the orthogonal direction (Y) is about 5 times smaller and can be ignored. Typically for these systems the bolt-up error will exceed the tilt in the Y direction mentioned. Be sure to tilt (shim) the focuser such that it is pivoted about the eyepiece field stop or CCD location based on your investigation. This naturally requires some translation of the focuser base.

In the case of systems employing correctors near ½ diameter (all-spherical system) the field tilt in X and Y is about the same. The spot diagrams examples in Chapter 11 gives the X and Y tilt values in degrees for optimization of the MTF. When all is balanced out you may even be able to knife-edge detect field curvature too! For additional information see Section 2.2.4.

4.2.9 Distortion

Distortion is an aberration that can be ruinous in positional astronomy particularly if unsymmetrical in nature. The tilted component medial has no parent axis and has unsymmetrical distortion, but fortunately it has a very low value for the designs presented. For a detailed look at this aberration see Section 2.2.3.

Chapter 5
The Standard Medial

5.1 The Baseline System

Reducing this complex telescope to a simple, compact form suitable for amateur construction was a priority for all those involved in initiating the Schupmann Club. It became clear that Ed Olson's pioneering use of a tilted field mirror, rather than the usual field lens, offered the best chance of success, thus defining the baseline system or configuration. Although James Baker independently proposed the field mirror scheme, the author found, after careful reading of Schupmann's 1899 U.S. patent, that Schupmann was ahead of us all. On page 3, sections 70 through 95 of the patent, he introduces the idea of positioning the correcting device, not interposed as in the brachy, but *after* the objective's first focus and concedes that the secondary spectrum is still bothersome. We read

> thereby we lose one part of the advantage with respect to secondary spectrum, but we can regain it by placing in the focus of the object-glass a convex lens or a small *concave mirror* or a prism, which latter has at least one convex surface... .

Schupmann goes on to explain the action of the field element to bring about near perfect color correction. Moving the corrector beyond focus and introducing a field lens are, of course, his two breakthrough ideas, which clearly distinguish the brachymedial from the medial.

As detailed in Chapter 3, adopting the field mirror scheme for the baseline configuration constrained the design layout (neglecting folding) to mostly variations in the corrector-to-objective diameter ratios. The crown glass corrector's optical performance is somewhat thermally sensitive, especially so during figuring, because of its in-glass reflection and its relatively high coefficient of expansion. One would like to make the corrector as small as possible to greatly reduce its thermal mass; however, various optical aberrations limit this direction. These aberrations can only be overcome with multi-element correctors. The best solution from a theoretical standpoint is to make the corrector as large as the objective. So, you see, the baseline system is really a family of corrector diameter possibilities, somewhat limited by *temporary* thermal problems at one end and unwelcome aberrations at the other. In the early days we promoted a standard or baseline design utilizing a half-diameter corrector and since the systems were intended for solar work, the main elements were separately figured to remove spherical aberration. The idea to look for an all-spherical solution within the baseline design family

Table 5.1
Prescription for a 10-inch *f*/10 Super Schupmann (All dimensions in inches)

Element	Glass	Radius 1	Radius 2	Thickness	Diameter	Air Space
1	BK7	56.648 CX	587.61 CX	1.0	10.0 CA	99.40
2	Pyrex	70.81 CC	Grind flat	0.30	1.0 CA	54.59
3	BK7	14.851 CC	29.055 CX	0.50	5.459 CA	54.59

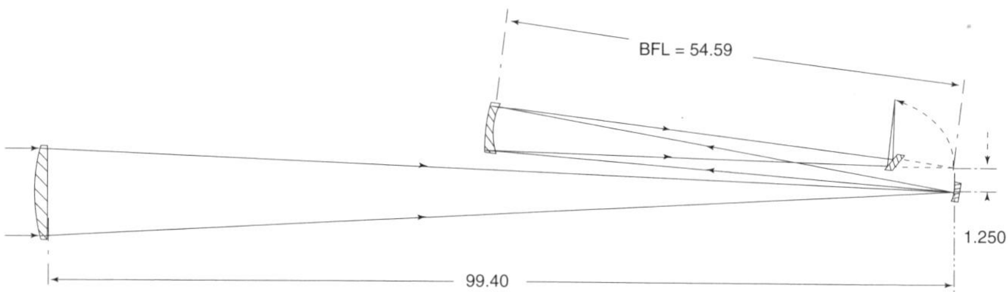

Fig. 5.1 *Optical diagram of a 10-inch f/10 Super Schupmann.*

never crossed our minds, so intent were we to have a sharp (first) focus at the solar occulting cone.

5.2 The Super Schupmann

In September 1969 I received a letter from Schupmann Club member, Clifton Ashcraft. Clif took a more generalized view of the Standard Design and, as part of his computer ray tracing work of the system, varied the corrector diameter (all the while controlling the radii for color correction) leaving all surfaces spherical. He discovered that spherical aberration went to zero when the corrector reached 54.6% of the objective diameter; or more exactly, when EFL/BFL is equal to 1.83, where EFL and BFL are, respectively, the objective focal length and corrector conjugate distance when working at 1:1 magnification (see Figure 4.1). Clif called this all-spherical solution "the" Super Schupmann and the name stuck. Again, you see, any medial employing a field mirror is part of a family we call the baseline system with the Super Schupmann a special, all-spherical, member.

Happily, especially for moderate size (8- to 14-inch) Schupmanns, this discovery gave us a corrector diameter that's a nice compromise, balancing the desire for good aberration control on one hand and easily managing thermal effects on the other. Various, non-singlet, corrector designs completely remove the thermal constraints and simultaneously open another degree of optical freedom (bending), but complicate the design from a builder's standpoint. These correctors are discussed in Chapter 8.

The Super Schupmann optical diagram, shown in Figure 5.1, is drawn more or less to scale and depicts a 10-inch *f*/10 system. The diffraction-limited field of view is approximately 0.15 degree, but opens out by a factor of approximately two by scaling the design to *f*/15. See Chapter 4 for curves of optical performance.

Table 5.1 gives the prescription for the all-spherical Super Schupmann. Radii and spacing tolerances are discussed in Chapter 11. Small errors in construction are easily compensated by slightly varying the spacing. Be careful not to exceed 0.77 degree of corrector tilt, which gives the 1.25-inch focal separations shown. Field center astigmatism is removed by slightly tilting the objective orthogonal to the corrector tilt.

5.3 Instrument Weight

What will a typical Schupmann weigh? The author's 7.5-inch *f*/13.3 standard baseline field mirror design, which uses a ½ diameter corrector, weighed in at 35 pounds. It was constructed with ¼-inch thick mahogany ply with aluminum cells along the lines described in Chapter 11. Maintaining proportionality, telescope weights generally vary as the cube of the aperture. We can expect a 10-inch to weigh about 86 lbs. The Stellafane medial, a 13-inch Super Schupmann, weighed in at slightly over 300 lbs., which is somewhat over the predicted 212 lbs. The increased weight was a result of using extra-thick ply skin to maximize the tube's stiffness and a very nice, rugged, double-ported tailpiece. Since this instrument is permanently mounted the trade-off between stiffness and portability was not an issue.

5.4 Light Transmission

Neglecting reflection loss, normal optical crown glasses exhibit small, but measurable internal transmission loss. This internal loss for BK7 is less than 1% per inch thickness in the C to F region of the visible spectrum; however, as the violet region is approached these absorption losses rise steeply, reaching 10% at 340 nm and 97% per inch at 310 nm. A 10-inch Super Schupmann has a total glass thickness of about 2 inches, leading to approximately 1% internal loss in the visible and about 5% at U-band centered at 370 nm. Thinning the objective to 0.75-inches can be safely done to improve transmission in the blue region for the 10-inch design; only a small spacing change is required.

Reflection losses through the visible average 4.25% per refracting surface and, after including the small internal loss, a conservative value of 90% transmission can be assigned to each lens element giving 81% transmission for the two lens elements of the standard design. The internal reflection at the glass-silver interface of the corrector is very high and will be neglected in this estimate. The field mirror will reflect about 90% for ordinary evaporated aluminum and much higher for special enhanced coatings. The resulting total system transmission for visible and near-IR transmission is 72%, about 2% greater than an equivalent standard crown-flint uncoated achromat. The optional diagonal mirror will reduce these values further. The deep violet throughput is typically far less, probably no better than

Table 5.2 Prescription for 4-inch *f*/12 Introductory Medial

Element	Glass	Radius 1	Radius 2	Thickness	Diameter	Air Space
1	BK7	27.133 CX	302.4 CX	0.500	4.0 CA***	47.638
2	Pyrex	48.248* CC	Grind flat	~0.25	1.00 CA	47.929
3	BK7	19.500 CC	32.500** CX	0.500	4.0 CA	48.08-Focus

* Aluminized surface.
** Silvered surface.
***The lens blanks should be at least 1/4-inch larger in diameter.

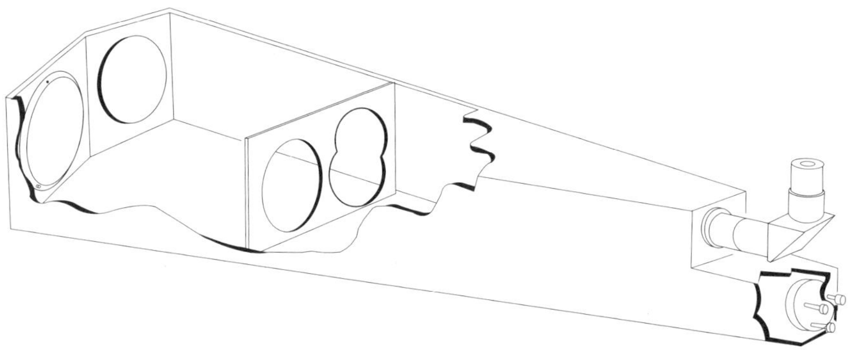

Fig. 5.2 *Cutaway view of 4-inch f/12 Introductory Medial.*

60% integrated across U-band. The visible transmission estimate is very conservative and 75% should be achievable by specially coating the field mirror. Not much can be done about the U-band performance short of special UV glasses or quartz lens elements. As a practical matter, most amateur CCDs have very poor U-band response, putting this spectral region almost out of reach anyway.

Antireflection coatings are not advised for the medial; however, the refracting, double-pass, surface of the corrector could be coated if the diameter is not much over 6 or 7 inches and extreme care is taken not to overheat the element in the vacuum chamber. This will markedly increase the system transmission, but an expensive broadband coating is still required for CCD work.

Chapters 10 and 11 provide the designer and builder more details on this design as well as important information about the layout and other arrangements. Next, the full-aperture corrector version of the baseline system is presented by way of two designs: an introductory 4-inch scope and a rather sophisticated 6-inch quartz example.

5.5 A 4-inch *f*/12 Introductory Medial

In 1964, Edwin Olson designed this instrument to introduce the medial to the ATM community. The design appeared in *Schupmann Club Bulletin*, issue No.1,

1966. Two successful instruments were made, one of which is now used by Bert Willard for prominence observing. A photo of Bert's solar medial appears in Figure 12.9. This design was featured in a technical report[1] by noted designer Dick Buchroeder, and his ray trace analysis showed fine performance.

Basically, the design follows Olson's original 6-inch aperture full-size corrector system seen Figure 12.8, however, the focal ratio was reduced from *f*/15 to *f*/12. The reduced *f*/ratio leads to a nice compact portable scope. This design is unusual in that it employs an out-of-plane optical layout, that is, the corrector tilt is orthogonal to the field-mirror tilt. This made it possible for Olson to baffle the system with a series of circular "stops". He later discovered that a simple twisted rectangle of thin sheet metal, when placed between two bulkheads near the field mirror, completely sky-baffled the system. Although not used in other designs presented in this book, the out-of-plane idea can be applied to any tilted-component medial.

This is a delightful little scope to build. Because the corrector is full-size it is nearly free of spherical aberration with spherical surfaces and can be monochromatically null-figured, thus leaving only the objective to aspherize (conic constant = –0.733 on objective's first surface) to a total system null in autocollimation. With its simple tube made of 3⁄16-inch thick plywood, a lightweight scope is achieved. Table 5.2 gives the prescription and Figure 5.2 shows a sketch of the instrument.

Make the objective and corrector center-to-center distance 6-inches. The optical centerlines between the field mirror and axial focus (without star diagonal) should be no more than 1.5 inches apart.

5.6 The All-Quartz Medial

5.6.1 A Dream Scope

The author has often contemplated the ideal medial from a purely optical viewpoint. Ignoring mounting, observatory and instrument cost or any other barrier; what would it look like? First, it would employ a full-size corrector to reduce spherochromatism and axial color to a few thousandths of a mm over a wavelength range spanning 3,400–10,000 Ångstroms, and simultaneously maximize the scientifically useful diffraction-limited field that results from using small tilt angles and shallow curves. Second, its clear aperture would be 12 inches, more than adequate for most serious amateur CCD work on double stars and interesting variables. Third, the focal length would be 180 inches so that a 2x apochromatic Barlow would provide the ideal magnification for CCD astrometry of double stars. At the 180-inch focus one could perform photometry on wider pairs as well as certain variables. By attaching a focal reducer, the full field would fill a large amateur CCD for research requiring modest fields, such as photometric studies to

[1] Buchroeder, Richard A., "Design Examples of Tilted-Component Telescopes," Optical Science Center University of Arizona, *Technical Report No. 68*, May 1971, pp. 4, 43–45.

determine OB1 cluster membership. Fourth, the optics would employ an extra-low expansion optical material, thus making its 12-inch corrector totally practical. The designer has very few choices of such low-expansion material with optical grade quartz (fused silica) the only real contender. Since cost is no factor in contemplating this "dream scope", we would utilize the best back-illuminated CCD to extend photometric measures to U-band where the basic optics, neglecting reflection loss, are totally transparent.

5.6.2 Practical Thoughts

OK, let's come back to Earth and see what an affordable all-quartz medial would look like. As a model we can do no better than copy Edwin Olson's 6-inch *f*/15 design seen in Figure 12.8. Substituting optical grade quartz for the BK7 used in the original, we arrive at a design insensitive to thermal problems with ultra-stable focal length and high UV transmission. The optical performance is impeccable over any affordable amateur CCD chip size. The "glass" cost will no doubt run in the thousands, say $2,500, but try to find a 6-inch apochromat with anywhere near the performance detailed here. Nothing said here should be construed as a negative statement on the suitability of a BK7 version of this design. Olson's instrument was of the highest quality, but the thermal settling of the 6-inch corrector often delayed the beginning of super imaging to about 1 hour after sunset. Digressing slightly, all telescopes, to some degree, have this problem and it is often difficult to disentangle very slow tube currents and stratification from temporary optical element distortion. Large doors on the tube side at each end are always a good idea to quickly ventilate the tube air just after opening the dome. By aiming the tube vertically a strong chimney action occurs sweeping out the hot air. The doors are then closed for observing. This is common practice on giant European refractors.

5.6.3 The Design Prescription

This is a member of the Standard Design family, thus utilizing a field mirror which should also be made of a low-expansion material such as ULE. The 6-inch aperture is sufficient for many interesting observing projects, particularly those of a solar nature such as CCD granulation density counts and lifetime studies. Consistent with other design prescriptions throughout this book, the dimensions are in inches and each element is taken in turn as an optician would approach it. That is, each element is considered in isolation of the others, thus we see no R3, etc. Ray trace fans will, of course, re-label the curves, thicknesses and spacings in the standard optical designing sign convention. Table 5.3 gives the prescription for a six-inch quartz medial.

5.6.4 Design Characteristics

In this design the Hg (mercury green) line is focused on the field mirror. This gives the best field performance over the specified super-wide wavelength span. The field mirror diameter of 1 inch will provide a ½-inch (0.31 degree) fully illuminat-

Table 5.3 Prescription for 6-inch *f*/15 Quartz Medial

Element	Glass	Radius 1	Radius 2	Thickness	Diameter	Air space
1	Quartz	48.236 CX	305.05 CX	0.635	6.30-Blank	90.189
2	ULE*	90.787 CC	Grind Flat	0.375	1.0	90.189
3	Quartz	34.860 CC	60.850 CX**	0.635	6.30-Blank	90.835 **

*Reflecting Mirror.

**Silvered

*** Distance to the final focus from R1 of corrector.

ed field of view over the full color range. (Read section on the field mirror in Chapter 10.) The tilt of the corrector should not exceed 0.32 degree which gives a beam separation at the field mirror of 1.0 inch. This is more than enough for the diagonal mirror to pick off the converging beam a few inches before focus, thus allowing image access to most low-profile eyepiece or CCD holders. Compensating orthogonal tilt of the objective is about ½ that of the corrector and as with all medials, fine-tuned on the stars. The design final focus is arranged to lie behind the field mirror surface about 0.66 inch to help with back distance from the diagonal mirror. This also aids in straight-through knife-edge bench testing in white light autocollimation. Make the objective and corrector center-to-center distance 8 inches. Carefully drawn full-scale layouts will help tremendously in designing the stray-light baffles. Olson successfully employed an out-of-plane arrangement in his cleverly disposed baffles. A mini version of this scope operating at *f*/12 is shown in Figure 5.2 and gives a good idea what this instrument, its big brother, might look like.

This design requires the objective only be aspherized as the corrector has little spherical aberration residual when left spherical, a great advantage for getting a real smooth surface on the sensitive reflecting side of the corrector. The conic constant for aspherizing the objective's first surface is –0.748, thus it takes the form (but hardly the strength!) of a Schmidt corrector plate with the reflex point located at the 70.7% zone. This aspherization includes correcting the slight residual of the spherical corrector. The builder may wish to separately correct the objective in autocollimation using an Hg source and, again in Hg light, separately figuring the corrector at its working conjugates. The system should be checked in autocollimation in white light as detailed in Section 11.2. An 8-inch version is certainly within the budget of a few builders—simply scale-up the prescription dimensions by a factor of 1.333.

5.6.5 Optical Performance

As with other medials, this design is coma free (note that the different front-to-back radii ratio of the objective for quartz is roughly 6.3 to 1 whereas in designs

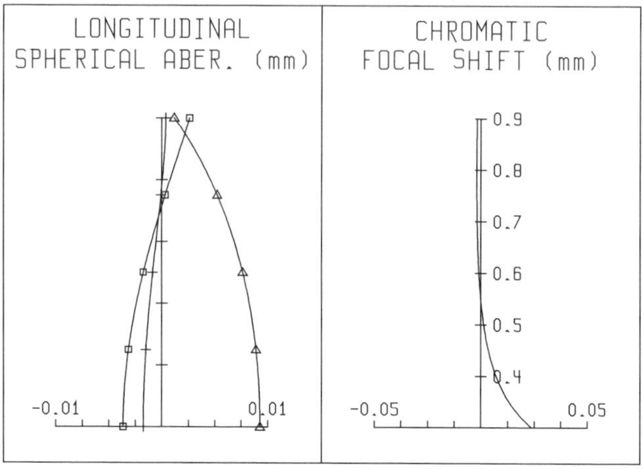

Fig. 5.3

employing BK7 it is closer to 10 to 1). The Petzval radius (curvature of field) is 18.14 inches and will be undetectable visually or in almost any CCD. The EFL is 91.208-inches yielding a focal ratio of 15.2, a focal ratio preferred by the writer for smaller instruments. Figure 5.3 is an OSLO trace of longitudinal color and color curve of the design. The graph of longitudinal color is in three wavelengths coded as follows: triangles = 365nm, crosses = 546nm and squares = 852nm. The horizontal scale is in mm and the vertical scale is heights on the aperture. We see here that the remaining spherochromatism is vanishingly small, undetectable by any known optical bench test. Looking now at the color curve in this figure we see the full impact of this instrument's superb color correction. The vertical scale is wavelength in microns and again the horizontal scale is in mm along the optical axis. To appreciate this level of correction one may look at the color curve of a standard Fraunhofer objective shown in Figure 9.4. This comparison, after normalization, shows this medial's "secondary" color to be over 100 times smaller in just the C to F region! It should be pointed out that this correction is not some magic property of quartz, as BK7 will, after adjusting the curves, give the same correction.

Chapter 6
The Split-Field Schupmann

6.1 Origins

All of Schupmann's medial designs, not just the Standard Medial, employ tilted optics to gain access to the focus. It is, nevertheless, possible to design a medial with un-tilted elements. The author has built such a system and found the design to be very practical, particularly in research where the inherently narrow field of the Schupmann is best applied. The geometric image quality is more uniform throughout the usable field compared with the standard medial, especially in the hybrid form to be described later.

James Baker proposed the original split-field scheme in his 1954 *Astrophysical Journal* article[1], "The Catadioptric Refractor," in which he argues the case for a specially designed medial for use in the most precise astrometric measurements of double stars and other demanding projects, as for instance, detection of the gravitational bending of starlight by Jupiter. The author can do no better than to quote Baker:

> In the split-field form the upper semi-circular half of the mathematical field is re-imaged by the Mangin onto the semi-circular lower half of the field at the second focus. The field stop can then be either a semicircular opening or a smaller circular opening within the semi-circle. Finally, one can use both the tilt of the Mangin and a partially split field to obtain the largest possible field angle for a given limit on image quality.

In the last sentence Baker suggests the hybrid system. The author designed his 9-inch split-field to leave the Mangin un-tilted and, instead, slightly tilted the objective to correct the tiny astigmatism at the field center. If you haven't guessed by now, the split-field design simply uses an off-axis portion of the mathematical full field of an on-axis, untilted component medial. A general idea of the scheme is shown in Figure 6.1.

6.2 Optical Layout

The split-field arrangement is best applied to medials employing a field lens, although a field mirror configuration is, with some difficulty, possible. The split-field medial using a field lens is quite long for its focal length unless a rather awk-

[1] See footnote number 8 on page 10.

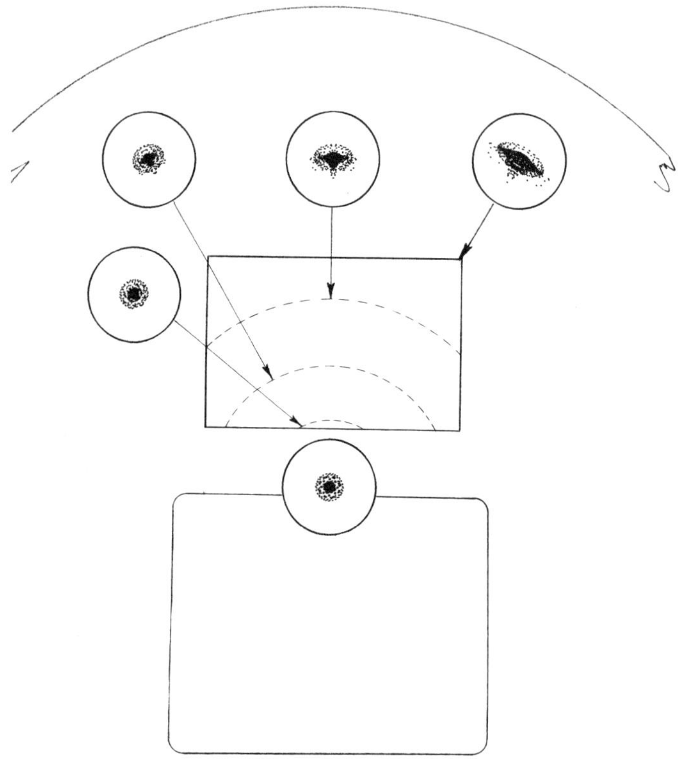

Fig. 6.1 *Schematic axial view (no pick-off mirror) of the focal region of a typical split-field medial. The partial arc at the top of the figure represents a 25mm diameter achromatic field lens rim. The lower rectangle is the field stop for the objective's uncorrected rays and is sized for that added blur. The upper rectangle represents a 6.9 x 4.6mm CCD chip, and where the rays passing through the field stop are imaged by the untilted corrector. The spots in this example are traced for an 8-inch f/12.5 system employing a 4-inch corrector. The wavelengths traced are 430, 580 and 850 nm. A circle surrounds each spot and is the size of the Airy disk. Arrows leading from each spot point to dotted arcs around the optical axis along which that specific spot shape is maintained. A spot centered over the field lens shows the axial performance, but is inaccessible. As always, the field lens is made generously oversize for ease of manufacturing.*

ward elbow is employed. In the field lens setup sky, baffling is virtually perfect and stray light uniquely controlled. The photo in Figure 6.2 shows the tailpiece section of my 9-inch experimental split-field medial. The quasi x – y positioning mechanism for the instrument's field lens is shown in Figure 6.3. Only the lower ½-inch diameter circular portion of the field lens is used. The diagram in Figure 6.4 shows the optical path within the ¼-inch plywood corrector housing. A photo of the 2.25-inch corrector in its cell is shown in Figure 6.5. This corrector required 20 fringes aspheric figuring to null. A pick-off mirror lying just behind the field lens sends the converging cone to focus 4-inches above the mirror base plate. Figure 6.6 shows the pick-off mirror and manually operated shutter paddle for taking

Fig. 6.2 *Tailpiece of author's 9-inch split-field medial.*

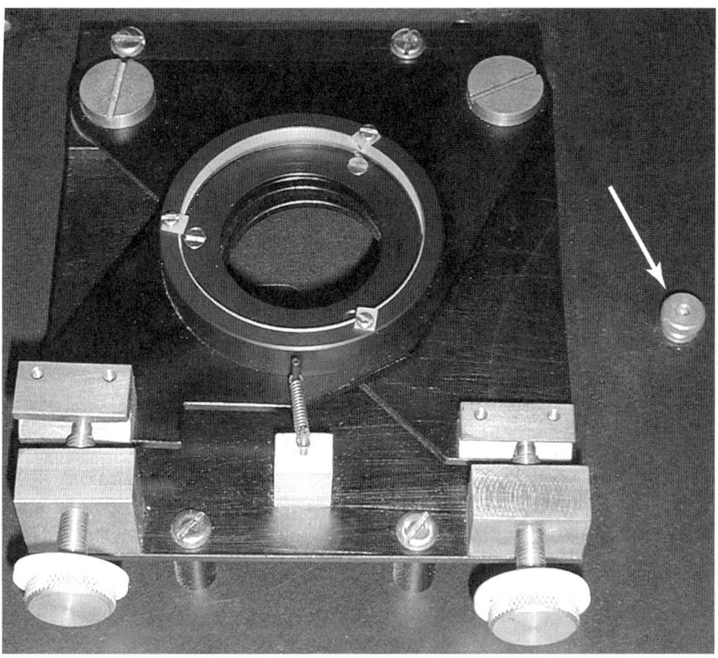

Fig. 6.3 *Field lens and X–Y stage for zeroing lateral color and atmospheric dispersion. A small knob (see arrow) is used for flipping in an alignment target located behind the plate.*

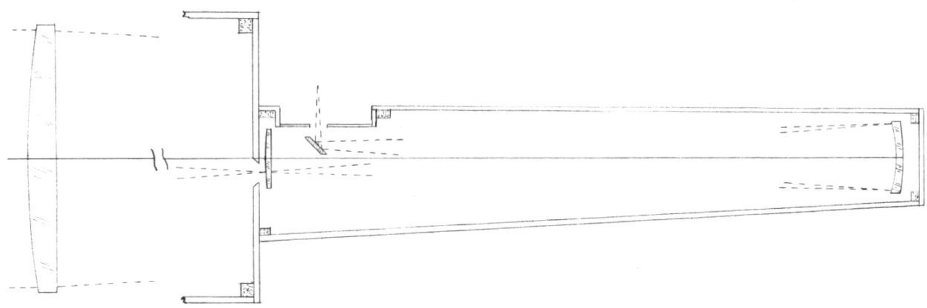

Fig. 6.4 *Schematic view of tailpiece tube and ray path of the author's 9-inch f/11.1 split-field medial. (note the tiny 2-inch corrector).*

Fig. 6.5 *The 2-inch corrector in its cell.*

CCD dark frames when using electronically shuttered CCDs. This beats blocking the aperture which can be as high as 12 feet above the observatory floor! The useful field is very small with this instrument, only about 10 minutes of arc, but far more than that required for the double star work for which it was designed.

A hybrid design for a split-field Schupmann has been tried by the author and the spot diagrams show that tilting the objective very slightly improves the overall field performance.

6.3 Willard's Anastigmatic Split-Field Schupmann

The dominant remaining aberration of the split-field design is, as with any medial, astigmatism. In March of 1990 Bert Willard was helping the author design an all-spherical split-field medial. This design utilized an achromatic field lens and a slightly separated mirror behind a single-element meniscus corrector lens. This, of

Fig. 6.6 *The pick-off mirror and cell along with manual dark-frame shutter of the 9-inch split-field.*

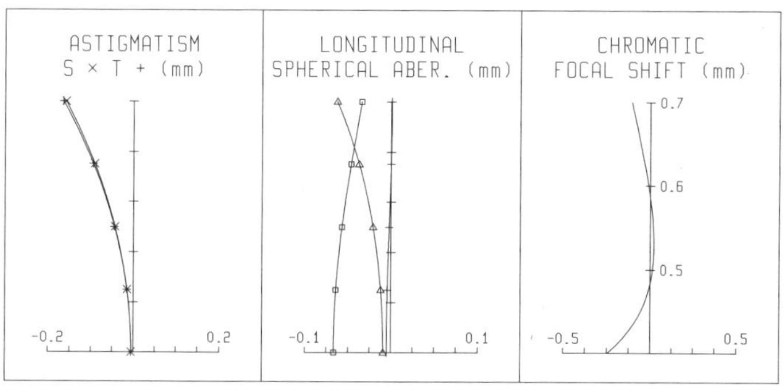

Fig. 6.7 *OSLO ray trace results for Bert Willard's anastigmatic 6-inch f/17 split-field medial.*

course, places the corrector in double-pass and allowed a spherical solution with a 40%-diameter corrector element. Bert noticed that most of the astigmatism was coming from the mirror element and that by moving the mirror further from the corrector, while holding the EFL constant, the signs of the astigmatism could be made to add up to zero! The mirror diameter grows to nearly that of the objective, reminiscent of the full-size corrector systems! *This represents a fundamentally new development in the design of the medial form since its invention.* The design lends itself well to solar tower installations where the extra instrument length is easily accommodated. If the amateur has a large observatory and wants the best astrometric instrument (distortion only 0.001%) the design is perfect, but a tall equatorial mount will be needed. Astigmatism, spherochromatism and color curve of a Willard anastigmatic split-field medial are presented in Figure 6.7. The wavelengths used for tracing are C, d and F. Note the astigmatic curves lie over one another which demonstrates a near perfect compensation for astigmatism, the curve now only indicating curvature of field—amazing!

Table 6.1
Prescription for Willard 6-inch *f*/17 Anastigmatic Split-field Medial

Element	Glass	Radius 1	Radius 2	Thickness	Diameter	Air space
1	BK7	56.644 CX	570.60 CX	1.0*	6.0 CA	99.330
2	BK7	36.73 CX	36.73 CX	0.236	1.250	40.862
3	BK7	11.515 CC	34.057 CX	0.472	2.726 CA	23.622
4	ULE**	42.653 CC	Grind flat	~0.750	5.590	41.417***

* Thickness for 9-inch maximum aperture.
** Reflecting mirror.
*** This is the distance from R_1 of the corrector to focus.

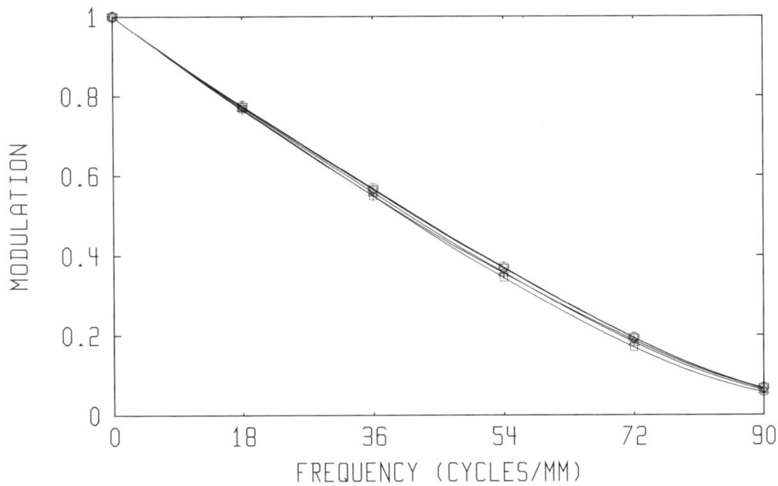

Fig. 6.8 *MTF curve for Willard 6-inch f/17 anastigmatic split-field medial.*

Table 6.1 gives the prescription for an all-spherical 6-inch, *f*/17 Willard anastigmatic medial.

The design presented here is simplified by the use of a singlet field lens. As a result, slight secondary lateral color remains. For critical work over the full solar disk it is advised that the builder achromatize the field lens. This entirely eliminates secondary lateral color and also gives slightly better geometric axial correction. Antireflection coating of the field lens' outer surfaces should be considered; the matching radii inner surfaces are, in any event, cemented. Thinning the objective is entirely practical for both designs presented, as only a small spacing change between the objective and field lens is required to restore full performance.

Table 6.2 provides the expert builder all he needs to complete the optics of a fine 8-inch *f*/12.5 Willard anastigmatic split-field medial. This faster design features an achromatized field lens.

In these designs, as with any split field instrument, the curved focal surface can be flattened—at the expense of introducing strong distortion—with a simple

Table 6.2
Prescription for Willard 8-inch *f*/12.5 Anastigmatic Split-field Medial

Element	Glass	Radius 1	Radius 2	Thickness	Diameter	Air space
1	BK7	56.644 CX	587.60 CX	1.0	8.0 CA	99.685
2	BK7	21.970 CX	14.947 CX	0.315	1.770 CA*	Cement
3	SFS2	14.947 CC	48.130 CX	0.236	1.770 CA	40.862
4	BK7	11.515 CC	34.057 CX	0.500	3.807 CA	23.622
5	ULE	42.653 CC	Grind flat	~1.00	7.540 CA	40.547

* Field lens sized for full ½-degree field, make it 2.0-inch OD. Remember that only a small circular (0.880-inch dia.) off-center patch is actually utilized. The lens must be made full size and wedge-free even though little of it is used. A cored-plug lens could be implemented, however alignment would be close to impossible! Also, since the diagonal comes well before the field lens, minimizing the lens diameter serves no practical purpose. Provide handy *x* and *y* adjustment of the field lens as described in Figure 6.3.

plano-convex lens near the focus. Alternatively, the CCD or eyepiece can simply be tilted about the off-axis field point selected for the field center giving a best-fit plane to the spherical 16.4-inch radius field curvature. The best focus will occur where the plane intersects the 70.7% field point. This solution, in the case of the Willard anastigmatic, will give a full ½° of diffraction-limited performance in the split-field configuration. The MTF curve in Figure 6.8 shows the best focus system performance of the 6-inch *f*/17 Willard anastigmatic over a 0.25-degree untilted FOV. Tilting the CCD or eyepiece just 20 minutes of arc will give performance indicative of a theoretically perfect MTF. The 8-inch *f*/12.5 version will cover a full ½-degree FOV with the same high performance, and the best compromise CCD tilt will be about 55 minutes of arc.

Other split-field designs employing a close-coupled mirror element or a single element back-silvered corrector will, of course, have smaller diffraction-limited fields.

A split-field medial is not recommended unless the builder has a permanent observatory. For those who require a narrow-field instrument with a very low level of distortion and high definition, a split-field may be just the thing.

The CCD photo in Figure 6.9 shows the triple star Zeta Cancri taken with the author's 8.3-inch *f*/12 split-field Schupmann. The focal length was increased to 301 inches with a 3x Barlow. Note that the close AB components are easily resolved in spite of their sub-arcsecond separation.

Figure 6.10 is an isometric view of the 6-inch *f*/17 Willard anastigmatic medial. Details of the focal region which include the pick-off mirror are not shown.

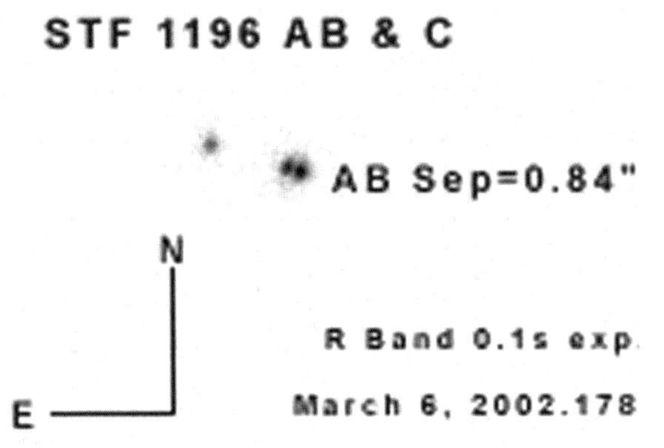

Fig. 6.9 *CCD image of Zeta Cancri taken with the author's 9-inch split-field medial.*

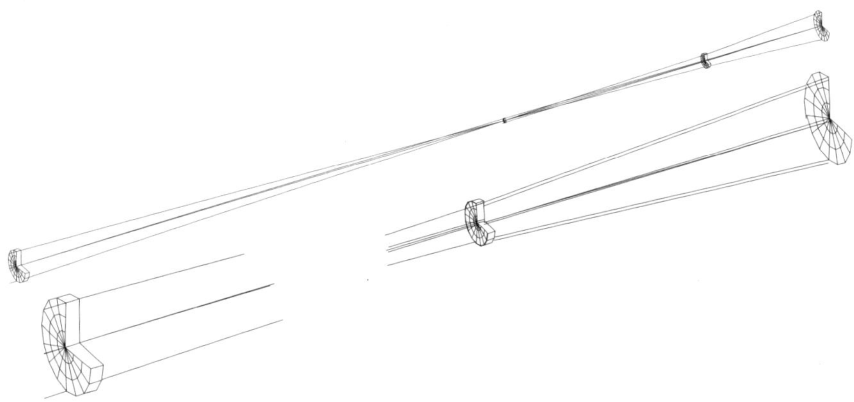

Fig. 6.10 *Isometric view of Willard anastigmatic split-field medial. The widely-spaced reflecting element is seen at the far right.*

Chapter 7
The Solar Schupmann

7.1 Introduction

Of the hundreds of "white light" telescope designs only the Schupmann medial is coronagraphic by nature. This unique characteristic allows the observer to measure a double star in the evening and observe solar prominence detail the next morning. We need only flip in an occulting disk or cone over the field mirror and a Lyot stop over the corrector, and with a simple #92 Kodak or Schott #RG645 red filter in the eyepieçe, prominences are beautifully revealed. (An ambitious amateur can fit his medial with a field-mirror turret to quickly switch between planetary, white light solar and prominence work.) The sky must be very clear for this to work well, so we usually employ an inexpensive interference filter centered on the Hα line at 656.3 nm. Using a narrower bandwidth filter increases the contrast by reducing the background sky intensity. A filter width of 10 nm (100 Å) at half power is just fine.

Yes, you can purchase either a special narrow-band (sub-Ångström) Fabry-Perot etalon tailpiece filter or a similar aperture filter costing thousands of dollars and simultaneously observe the prominences and surface features in Hα light. This sort of device is limited to around 3 or 4 inches due to the rapidly increasing cost of required full-aperture pre-filters. For the Schupmann user, however, the joy of observing solar prominences with a simple inexpensive homemade instrument is hard to surpass.

7.2 Theory of Operation

This discussion is specific to the observation of solar prominences. Other than a few special projects, such as searching for faint moons around bright planets and measuring earthshine, the coronagraphic properties of the Schupmann are of little benefit to nighttime observations. Observing the Sun's prominences and inner corona place special requirements on a telescopic system. The prominences themselves are the brightest visible emission features and mark the head of the famous hydrogen Balmer series at 6562.8 Å. Over a very narrow optical bandwidth, their brightness is comparable to the photosphere. It would seem that simply blocking the sun with a distant mask (the corner of a tall building) one could at least explore a portion of the solar limb for prominence activity with little optical aid. Unfortunately, the sky brightness at the solar limb reaches its maximum and completely overwhelms any prominences. Additionally, in this simple example, light diffracted by the building's edge further adds to the field intensity.

Let us now address the problem of observing prominences with a single element objective. The Sun's image is first occulted, for instance, with a flat circular reflecting mask located precisely at the Hα focus. This gives us an artificial eclipse. The mask diameter is calculated to be slightly larger than the solar image. We may wish to tilt the mask slightly to send the intense reflected beam out a tiny window and into an optical absorber. It would seem at this point that all we need to do is to carefully observe the focal region with a low-power eyepiece equipped with a deep red filter. To our dismay no prominences can be seen. Even with a moderately narrow-band interference filter our chance of seeing one would be slight. Very bright "scattered" light floods the field surrounding the occulter. Puzzling over this, we remove the eyepiece and after being sure the Sun is still centered, gingerly peer past the mask edge through a red filter. There it is—the edge of the objective is a brilliant, almost blinding, ring of diffracted light. On the lens surface every tiny sleek, dust speck and smudge is painfully apparent. We also observe a bright image of the sun hanging in the space near, and centered over, the objective lens. It is this extraneous light that's flooding the focal region surrounding the occulter, substantially adding to the sky brightness.

A French solar astronomer named Bernard Lyot wanted to build a telescope to study the inner corona at the green 530.3 and red 637.4 nm iron emission lines without a natural eclipse, thus providing a day-to-day record of these phenomena. The success of his "Lyot Coronagraph" was the result of his intense study of every possible source of harmful instrumental light. Lyot's great breakthrough was the introduction of a Fabry or field lens just after the occulter, and a relay lens located just after the image of the objective formed by the field lens. Now the bright diffraction ring could be blocked with a suitably sized aperture stop centered on the relay lens, thus securing a much darker field around the relayed image of the occulting mask. The bright ghost image of the Sun was similarly blocked with a black spot painted on the center of the relay lens. Second in importance, this source of harmful light comes from a double reflection between the objective's first and second surfaces. It lies about ⅙ of the primary focal distance, thus the Fabry lens images it on the relay slightly defocused requiring a somewhat larger blocking patch. Lyot delved into every possible additional source of scattered light, focusing primarily on the objective's internal and surface scatter. For his work the glass had to be made especially free of bubbles. Great attention was placed on the quality of the polish and its maintenance in the harsh, 9,400-foot altitude and conditions of the Pic du Midi Observatory in the French Pyrenees, for it was here that the required coronal skies prevailed and where Lyot, at last, successfully photographed the inner corona under a blazing daytime sky. He isolated the coronal line features with relatively wide-band polarizing birefringent filters of his own devising.

Lyot probably never knew of the Schupmann medial arrangement, which is, ignoring the occulter and Lyot stops, fundamentally indistinguishable from Lyot's

Table 7.1 Prescription for a Field Mirror Type

Element #	Glass	Radius 1	Radius 2	Thickness	Diameter	Air space
1	BK7	57.126CX	404.61 CX	0.551	6.50 CA	97.047
2	ULE	70.066CC	Grind flat	~0.5	1.25 CA	54.622
3	BK7	15.160CC	29.449 CX	0.315	3.664CA*	56.197-Focus

*Lyot stopped to 3.36 for prominence work.

classic coronagraph or prominence scope. Taken point by point: the medial has a singlet objective, which can be worked to the best low-scatter standards; It has a field lens (or mirror), which images the objective onto a relay element—the corrector; just add the occulter, Lyot stop (s) and appropriate filter and—presto!—you have a coronagraph, a most delightful instrument for viewing the prominences in Hα light.

7.3 Practical Designs

Unless you have a favorable site, daytime seeing conditions usually limit solar detail (resolution) to about 1 arcsecond. Attaining even this level of seeing requires that the astronomer find the best time of day to observe. Usually the morning Sun is best, with a second good period of seeing about two hours before sunset. The design presented here is, as with all such things, a compromise. For the moments of best seeing we would love a 12-inch aperture, but who could afford it. A 4-inch medial would qualify as a serious solar instrument and the design detailed here could be scaled to that size. The author feels that a 6.5-inch *f*/15 is the best of both worlds; not too expensive to build and mount and yet offering exciting solar resolution. The 1 arcsecond requirement is easily met and its nighttime usefulness is not seriously compromised for many other areas of amateur research.

The design is based on the all-spherical Standard Medial with only this small modification: the field mirror (or lens) is positioned just 0.080-inches beyond the objective's Hα focus and the curves adjusted for better imaging properties at the solar limb without compromising the on-axis performance. Smaller corrector-to-objective diameter ratios can be employed; however, for reasonable solar rim imaging the focal ratio would need to be drastically increased. The design is Lyot-stopped to *f*/16.6 for sharp prominence imaging. At this *f*/ratio the objective need not be separately corrected for spherical for a sharper first focus. With all spherical surfaces only a slight touch-up is needed in the autocollimation phase. The builder should include a 2-inch focuser to better view the full diameter solar image at low power as well as an eccentrically rotatable 1.25-inch eyepiece (or CCD) set-up to scan the limb at high power. The author has designed a field lens system for those who would prefer the traditional Lyot coronagraph-like layout. By using Kutter's layout, the instrument length remains about the same as for the field mirror arrangement.

The occulting cone diameter should be slightly larger than the winter Sun's

Table 7.2 Prescription For a Field Lens Type

Element #	Glass	Radius 1	Radius 2	Thickness	Diameter	Air space
1	BK7	57.126 CX	404.61 CX	0.551	6.50 CA	97.500
2	BK7	18.475 CX	Grind flat	0.157	1.25 CA	54.606
3	BK7	15.113 CC	29.429 CX	0.315	3.677 CA*	56.063-Focus

*As per field mirror arrangement.

image, which is 0.908-inches for this prescription. This allows for tracking inaccuracy, etc. Normally the telescope pointing is accomplished with the slow motion controls to bring the limb to almost peek out from behind the occulter. So, an oversize cone is soon greatly appreciated and the slight mismatch in radii is of no consequence when studying *individual* features at high power. For full-limb imaging the builder will want three occultor diameters, 1) Spring and Fall, 2) Winter, 3) Summer.

Both designs (shown in Tables 7.1 and 7.2, are of identical 100.18-inch focal length. The corrector is set up to magnify slightly which brings the final focus about 1.6-inches beyond the vertex of the field element. This gives better access to the focus by bringing its position well outside the tube, after reflecting from a diagonal mirror for instance. The tilts should be minimized such that the beam separation at the field element is no more than 1.5 inches.

7.4 Field Mirror — Interchangeable Set-ups

If the builder decides to have interchangeable field elements for each observing situation, the following ideas may be helpful. Let us consider the field mirror arrangement first. Three plug-in mirrors will need to be made, one (aluminized) of ULE holding the occulter, a second (uncoated) of optical grade quartz for white-light surface observations and a third, aluminized, for night work. For the prescriptions in Tables 7.1 and 7.2 the location of the polished occulter rim should be no more than 0.080 inches in front of the field mirror. Just enough to blow away dust is plenty of space. Greater spacing is not possible in the field mirror arrangement due to down-and back-beam clipping from the tilted element. The occulter can be cemented carefully in place with a soft RTV-type adhesive using a centering jig. A hollowed area is machined on the center of the cone's back to contain the cement such that a hard circular edge contacts the mirror surface. Alternatively, the cone could be bolted through a central perforation in the field mirror.

By far the best idea, from many standpoints, is the "occulting hole" field mirror (see Figure 7.1). The hole is cored to nearly the same size as an occulting cone. After this roughing, the hole's inside diameter is spindle-ground to exact size with a slight (~4°) taper, with the largest diameter at the back to allow for tilt and expanding rays. Special care is required to get a perfect chip-free edge at the mirror side. This is all easy work for any competent professional optician and could be farmed out to a local shop for diamond generating. In use, a blackened light trap

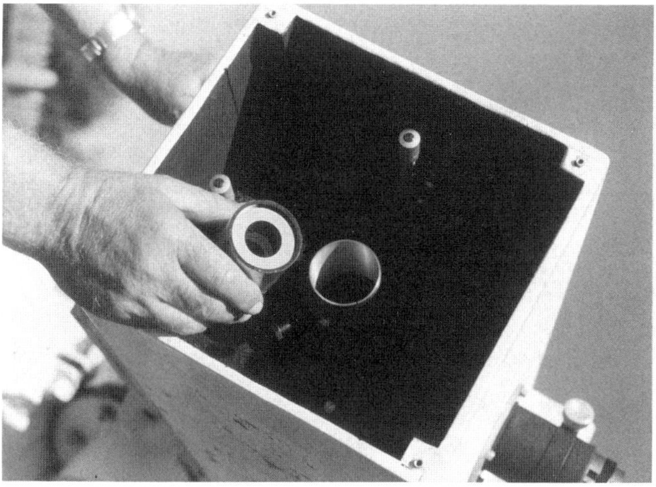

Fig. 7.1 *Larry Prugnarola's "occulting hole" plug-in field mirror. (Note the wide stance of the micrometer barrels for fine tilt control.)*

is advisable to contain the emerging (dangerous) beam. It will need plenty of grazing incidence surface area to keep hot spots from developing. A tilted tin can works great!

For the white-light solar field mirror, use optical grade quartz wedged to reflect the back reflection back up the tube to the side opposite the corrector. Send the transmitted beam into an optical absorber as above. Further light reduction is required at the eyepiece or serious eye damage may occur. Finally, the nighttime field mirror is simply aluminized.

7.5 Field Mirror Turret

Much neater than the plug-ins described above, the turret arrangement requires only one oversize field mirror. It is made from optical grade quartz with all but a 120° segment aluminized. The mirror is wedged to send the reflection from the polished second surface to an optically black absorber. It can, of course, utilize either an occulting hole or a small-cored hole for bolt-through interchangeable cones. The adjustable mirror cell would hold a sub-cell, which incorporates a central spindle. Three 120°-disposed detents position the mirror for the three functions. The spindle axis must be located on a line passing through the first and final focus. This minimizes the corrector tilt for a straight through view. If using a diagonal mirror the spindle axis location is unimportant.

7.6 Flip-In Occulter

If one is willing to use a solar aperture filter for surface work, then one aluminized

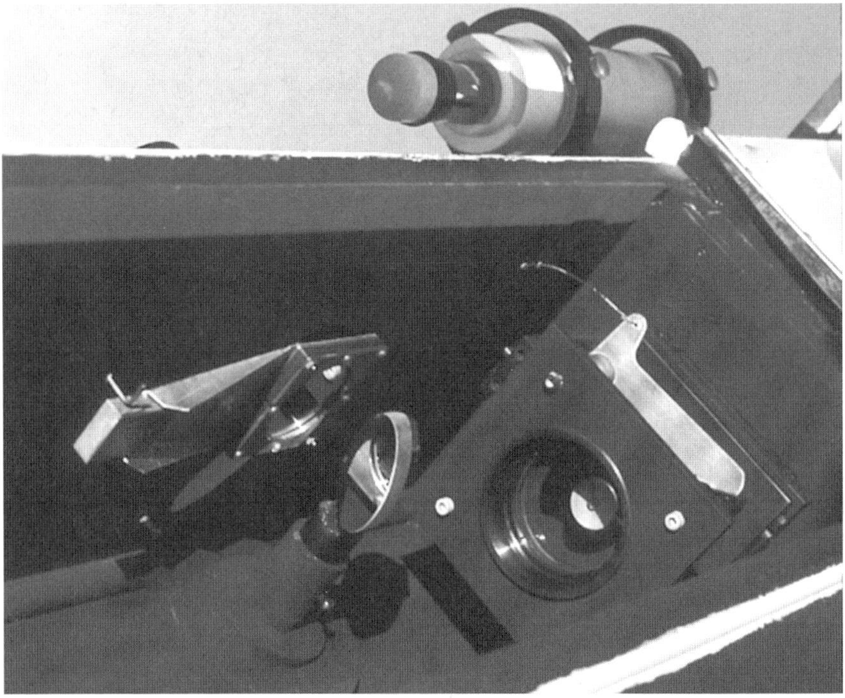

Fig. 7.2 *The flip-in occulter of my 7-inch solar/planetary medial can be seen to the right of the diagonal and through the field lens which is located near the bottom right-hand side of the image. The interference filter, to the left of the diagonal, is mounted on a pivot to swing it out of the path for night work.*

field mirror is all that's needed. The occulter is supported by a thin, short length of stiff piano wire and simply rotated or pivoted into place just shy of the mirror surface. Flip it away and you're ready for the evening's observations! Position the wire near one of the solar poles to lessen the chance of it covering a prominence.

7.7 Field Lens Arrangements

The field lens design prescription shown in Table 7.2 can be used for an in-line configuration or in, lets say, a Kutter arrangement. For the field lens design the correct position for the cone is 0.580-inches in front of the lens element, and it can have an integral center stalk that is bolted through a centrally-drilled hole in the lens. The design keeps dust specks well away from focus because of the greater spacing. Better still is a flip-in occulter as outlined above for the field mirror. The author has used such an arrangement with his 7.5-inch *f*/13.3 in-line system with excellent results. (See photo in **Figure 7.2** which shows the wire-supported flip-in occulting disk, field lens, diagonal mirror and—shown rotated into position—interference and blocking filter.)

7.8 The Occulter

Cone shaped occulters distribute the Sun's heat and light evenly around the inside of the tube. A special ring of black-painted copper sheet could be placed to intercept this energy, however it would soon heat enough to cause tube currents. By soldering a loop of soft 3⁄16-inch OD copper tube to the outside of the ring, and running ambient temperature water through it, the heat will be almost totally removed. A small fish tank pump, a bucket of water and some plastic tubing complete the package.

Flat-disk occulters are usually tilted slightly to aim the Sun's energy to a water-cooled black plate located just outside the incoming converging beam. Another scheme uses a window in the side of the tube to allow the energy to exit the tube. Sending the beam back out through the objective works too, but there will be some reflections that may reduce contrast slightly. Excessive tilt angles in flat-plate occulting disks will cause opposite edges of the occultor to become de-focused, and may cause it to appear slightly elliptical in the eyepiece, as well.

7.9 Polishing the Occulter

Cone-shape occulters are turned in a lathe to an included angle of 90 degrees (see Figure 7.3). A finishing tool will give a good preliminary surface for the polishing work. Fine #600 emery paper is cemented to a size block and held against the spinning cone, using honing oil for a lubricant. Keep the size block moving and in good contact until all traces of tool marks have disappeared. Find a piece of dry, hard maple and cut it to a roughly 1-inch square block. Flatten one end-grain side as best you can. This is your polishing tool surface. Coat the end-grain tool with "Simichrome Polish" a very high quality metal-polishing paste made by Happich of Germany. Before polishing the cone, first work the polisher on a flat piece of *clean* aluminum to break it in. Clean the polisher, apply more paste, and polish the spinning cone, using light pressure and lots of motion. Check progress by spin-wiping the cone with a soft paper towel. Don't attempt an "optical polish"—all that's required is a reasonably smooth and shiny surface.

Flat-disk occulters are finished as above, but not in the lathe. Make the disk about 1⁄8-inch thick so a 2-56 tapped hole for holding a support wire can be put on the back without breaking through. Finish by hand as in making a mirror. This polishing process takes about five minutes start to finish including, the #600 oilpaper lapping. Any TN capable of completing a Schupmann will have done something like this before.

In general the flat-disk type occulter adapts best to a flip-in arrangement, particularly with a field lens set-up; while the cone type is better suited to the field mirror arrangement, turret or plug-in.

7.10 Eyepiece View of the Occulter

One would think that a cone occulter in a field-mirror system, being brightly lit,

Fig. 7.3 *Bert Willard's occulting cone cemented to the field mirror of his 4-inch f/12 prominence medial.*

presents a major source of disturbing light in the eyepiece. After all, unlike a field-lens system, the corrector (relay element) is looking at the lit side. Amazingly, the eyepiece view of the occulter looks totally black when a simple deep-red eyepiece filter is used. This is because the 45° cone face is a very efficient reflector and a poor back-scatterer. The flat black absorbing surface surrounding the cone all but kills any possibility of returning light.

Getting the cone's sharp edge and the Sun's image at the same focus in Hα light takes a bit of fiddling with the field mirror position along the optical axis (more easily done with a plug-in set-up). This is safely done at night on the Moon or a bright planet using a #92 filter, and later refined on a prominence using your interference filter. In operation we naturally focus on the prominence, so a slight defocus of the occulter edge, although at first distracting, seems not to materially bother.

7.11 The Lyot Stop

To see the prominences really well you must include a Lyot stop aperture over the corrector, which for the designs discussed here measures 3.36-inch. It's amazing to see these delicate feathery features nearly, and sometimes completely, disappear when the stop is flipped out.

You may chose to make the Lyot stop permanent and thus lose ½-inch of aperture for night work, or you can arrange the stop to be pivoted via a handle on the outside of the tube that swings it in and out of position. The photo in Figure 7.4 shows the author's pivoted Lyot stop in position over the corrector. The pivot shaft can just be seen exiting the tube to the right; it is bent to form a flip handle. To

Fig. 7.4 *Flip-in Lyot stop for the author's 7-inch combination planetary-prominence medial.*

block the ⅙ objective ghost a small disk of black electrical tape may be permanently stuck to the center of the corrector's first surface. This tape spot only needs to be ⅕-inch in diameter, which is an effective ⅜-inch central obstruction and is of no real harm to the performance, although you will lose bragging rights about to owning a unobstructed telescope! Interestingly, Rush and Schnable suggested leaving a small central spot unsilvered on the corrector's back surface, thus letting the bright ghost through to hit a black absorber. Probably the best way, all round, is to incorporate a thin, 4-wire spider on the flip-in Lyot stop with the center spot glued at the wires' intersection. The author later used this method and it works just fine. The prominences are well seen without the central spot feature; it is mentioned here for those trying for the very best contrast.

7.12 The Schupmann as a Stellar Coronagraph

While on the subject of the uses of the coronagraphic properties of the medial, it is not entirely out of place in this solar chapter to describe the instrument's use as a stellar coronagraph. A "stellar coronagraph" does not, as its name implies, detect the corona of stars (at least not yet!). Its name is simply a carryover from the solar coronagraph, which is, as described earlier, an instrument that blocks the Sun's image and minimizes instrumental scatter in a wide region surrounding the Sun's limb. This "artificial eclipse" makes it possible to observe the inner corona under very clear mountain skies using filters centered on certain coronal lines.

Stellar coronagraphs are a rather new tool for the astronomer and are being used to discover and measure, for instance, circumstellar disks. A classic example of these possible proto-planetary disks surrounds Beta Pictoris, a nearby 3.8 magnitude A5 V star. Other nearby stars showing similar features include: Vega,

Fomalhaut and Epsilon Eridani, to name just a few. Stellar coronagraphs permit exacting studies of these disk features as well as faint stellar companions without the bothersome interference (scattered and direct light) of the parent star. As with a solar coronagraph, a mask is placed at the intermediate focus thereby blocking the unwanted light. Of course, in the case of a stellar coronagraph the mask is very tiny and in the best professional instruments allows working to within 6 arcseconds or less of the bright star.

Optically the stellar coronagraph embodies all of the principles set out by M. Bernard Lyot in his George Darwin Lecture[1] delivered on May12, 1939. Because the Schupmann design is also coronagraphic, certain stellar coronagraphic work can also be accomplished. As with Lyot's design, this work is limited to a fairly narrow spectral bandwidth because the intermediate focus is non-achromatic. Even so, by employing a comparatively wide stellar occulting mask very impressive CCD results have been achieved by the author. The parameters governing the effectiveness for a particular study are as follows:

1. Mask width (angle) subtended at the objective's focus: determines the maximum useful bandwidth of the filter (system light efficiency) and the sky area blocked.
2. Location along optical axis: selects the approximate mean wavelength of operation.
3. Filter bandwidth: sets the rejection ratio; i.e., intensity ratio of blocked light versus not blocked.
4. Spectral response of the CCD: this combined with the chosen bandwidth determines the sensitivity to faint stars, etc., lying outside the mask.
5. Focal ratio: sets the blocked spectral region for any specific mask width and location.

Schupmanns employing a field mirror can potentially extend the working bandwidth because the beam is doubly blocked; i.e., once before the field mirror and once again upon reflection. As the mask is moved away from the mirror face this bandwidth opens, however, this cannot be carried much more than 5mm due to the tilt of the field mirror itself. Vertical placement of the masking bar (in the direction of tilt) helps here, but is too restrictive for survey work.

As an example, the author's Schupmann stellar coronagraph utilizing a 0.5 mm black wire mask in near contact with the field mirror, gives a working bandwidth of about 70nm centered at 598nm. This particular setup allows looking for faint companions of nearby stars with separations as close as 20 seconds of arc. Exposures of 60 seconds solidly reach 12.5 magnitude making possible the detection of white dwarfs and main sequence stars as late as M5 for nearby bright stars. The rejection ratio is about 50 giving a good dark field for survey operation. Moving the masking bar away from the field mirror by about 5 mm, along with use of

[1] Lyot, Bernard "A Study of the Solar Corona and Prominence without Eclipses," *Monthly Notices of the Royal Astronomical Society*, Vol. 99, No. 8, June 1939, pp 580-594.

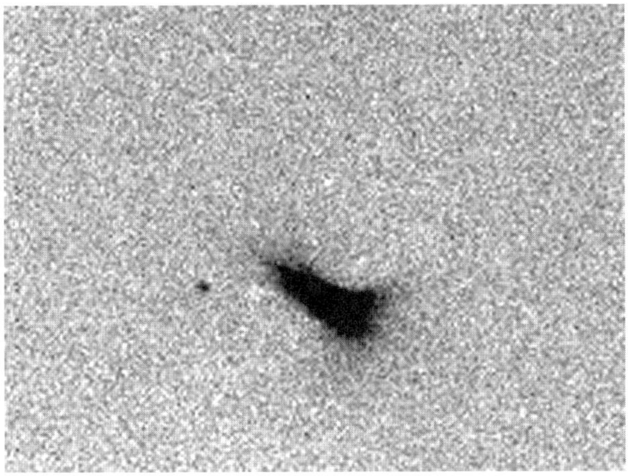

Fig. 7.5 *Altair 98% blocked revealing 12th magnitude companion. Unseen occulting bar runs from lower left to upper right.*

a broader filter, almost doubles the sensitivity. Currently (2005) the filters used are a photometric V-band together with a Schott OG 590. Special interference filters with steep shoulders will increase the throughput considerably. Figure 7.5 shows a new closer component (DAL 27AD) of Altair (STF 10), discovered by the author using the mask described above.

Amateur experiments may include imaging the faint moons around our favorite planets or finding faint components to nearby bright stars. The minds of advanced amateurs reading this have leapt well beyond the words here and are racing ahead with their own experimental ideas.

Mounting the occulting mask is most easily done by attaching it to the field mirror's swing cover as shown in Figure 7.6. The masking bar can be rotated to any desired position angle via an access port. This maximized the detectability of an attendant object by enabling it to be placed in the darkest part of the field. Figure 7.7 is a close-up of the rotatable "button" and its occulting bar.

For better control of the mask position, both laterally and along the optical axis, the builder is advised to install a tiny x, y and z stage. The stage's occultor support finger would be inserted through a port in the side of the tailpiece, and the stage itself dowel-pin indexed at bolt-up. Various masks can be imagined; the author has found that a slightly tilted reflecting glass sliver is close to optimum for many experiments. This limits the light scattered towards the corrector to about 1 millionth that of the occulted star, although even flat black paint on aluminum works well in practice. Tapering the mask would provide a means of adjusting the obscured width. Naturally a field lens system is simpler to instrument as scattering off the occultor impinges on the black tube walls and thus never finds a way to the CCD. A simple flat-black stellar occulting mask is recommended for a Schupmann using the field lens configuration.

Simpler, noncoronagraphic, occulting methods can also give extremely use-

Fig. 7.6 *Author's occulting bar held in place with the sprung wire secured by the upper screw. The little shouldered "button" holds the 0.5 mm wire occultor in a 3mm field opening. All are mounted on the swing-in dust cover.*

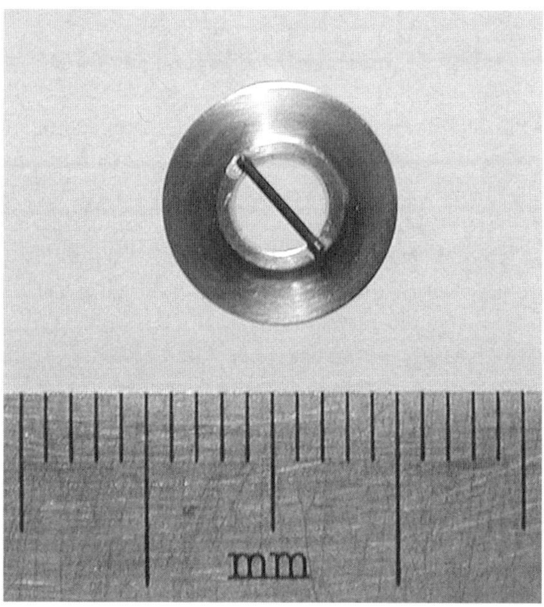

Fig. 7.7 *Rotatable "button" and occulting bar. This shouldered side faces the field mirror.*

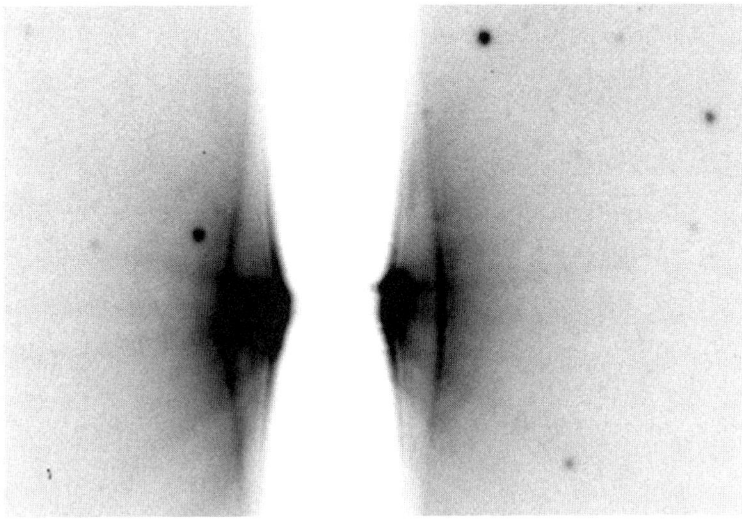

Fig. 7.8 *Polaris hidden behind a simple CCD window mask. Burnham's faint "C" (left) and "D" (top right) components are well revealed in this CCD image taken through the author's 9-inch instrument.*

ful results. One very efficient method is to place a mask directly on the CCD chamber window. Being located nearly at the final focus, the mask operates at a color-corrected location. The author used this method to recover and measure Burnham's faint 13th and 14th magnitude components (C and D) of Polaris[2]. See Figure 7.8 A CCD with the window close to the chip helps lessen the somewhat annoying diffraction caused by the mask edges. Of course, one could place the mask, usually an aluminum foil strip, directly against the CCD cover glass, but this placement involves the risk of damage to the delicate signal wires. Also, opening the windowed chamber gives a chance for dust to settle on the chip face. Although the coronagraphic properties of the medial are not utilized in this method, surprisingly good results are obtained.

[2] Daley, James A. CCD Imaging of STF 93 C&D, *Journal of Double Star Observations*, Vol. 2, No. 2, Summer 2006, pp 51–53. (A free internet journal: www.JDSO.org)

Chapter 8
The Type B Corrector

8.1 Introduction

Schupmann's patent drawing, shown in Figure 1.5, details four corrector configurations which vary from the simple singlet meniscus lens with the convex surface silvered, to the most complex type where two meniscus lenses are followed by a reflecting element. By dividing the refractive power among many surfaces, along with appropriate lens bending, Schupmann could correct axial color and spherical aberration in his giant medial designs, which employed these relatively tiny correctors. These ideas can be implemented, but at a price as the many glass air interfaces lead to substantial light loss. For instance, in his most complex corrector the light traverses eight glass-air interfaces and with the addition of the reflective surface yields a (corrector alone) transmission of about 62% without antireflection coatings. Also, reflections between the elements can give rise to ghost images in the final focus. Schupmann was aware of this and extensively studied the reflections of his preferred corrector type, "Fig 3" of the US patent drawing shown in Figure 1.5. He concluded, for that corrector, they were not harmful. Modern medial builders who wish to explore complex correctors will keep ghost image formation well in mind.

Most designs in this book employ a relatively large singlet corrector; however, the separated-mirror-single-lens corrector cannot be dismissed and shows definite advantages, particularly for medials of over 12-inch aperture. This type of corresponds to "Fig 2" in the patent drawing. We will call this corrector form Type B, which is also consistent with the convention Schupmann used in delineating the corrector designs for his brachymedials (see Figures 9.1 and 9.2). Advantages and disadvantages of this version are touched upon throughout this book and are delved into more deeply here. The following study is for BK7 lens elements only and other glass types will require separate analysis. Each objective is shaped for zero coma for all of the 3 examples shown.

8.2 The Main Advantages

1. By suitably bending the lens element, an all-spherical solution exists for medials employing correctors as small as 40% of the objective diameter. Curiously, a simultaneous solution for spherical and minimum axial col-

or exists near a 48% diameter and is discussed in Section 8.5. All these designs, at some bending strength, exhibit color correction well below detection, aspherized or all spherical (see 5, below).

2. The separated *mirror element* is always best figured spherical for any objective/corrector diameter ratio.
3. A mirror element of low-expansion glass, such as Zerodur or ULE, removes all but the slightest trace of thermal instability related to telescope operation. In addition, figuring the lens element in total system autocollimation is greatly facilitated with an already coated mirror element.
4. Protection of the reflecting surface is assured by rim-to-rim contact in a machined cell. In the case of a coronagraph system the mirror could be centrally cored to pass the ⅙ objective ghost, which would be absorbed in the black back wall of a sealed chamber, which also contains a screw-in desiccant plug. Now the mirror surface is truly protected.
5. Various aberrations including astigmatism are reduced compared to the singlet case. By suitable bending, spherochromatism can be made insignificant (comparable to a full-size singlet corrector) at the cost of slight aspheric work on the steep surface of the lens element. This aspheric work is in the form of an ellipse and is easy for any advanced ATM. When bent for an all-spherical solution, the spherochromatism is typically ½ that of a same diameter singlet corrector—in a special case described in Section 8.5—considerably smaller.
6. No in-glass reflection is present, thus relieving the lens element of figuring sensitivity.

8.3 Some Disadvantages

1. Two more glass-air interfaces reduce the transmission, making antireflection coatings almost mandatory.
2. Danger of ghost images or faint field flooding if the second surface curvature is allowed to be close to the mirrors'. This will not occur in a properly bent lens element.
3. Generally speaking, increasing the number of surfaces reduces contrast, however this is largely offset when considering the *effective* smoothness that can be obtained with no in-glass reflection present. Nevertheless, great care is required in polishing when dealing with this type if one is to match the scatter-free properties of the singlet corrector.

8.4 The Graphs

The first example, shown in Figure 8.1, is a design analysis aid for the author's 14.1-inch *f*/12.8 medial utilizing a ⅓ diameter (4.66-inch) Type B corrector, the

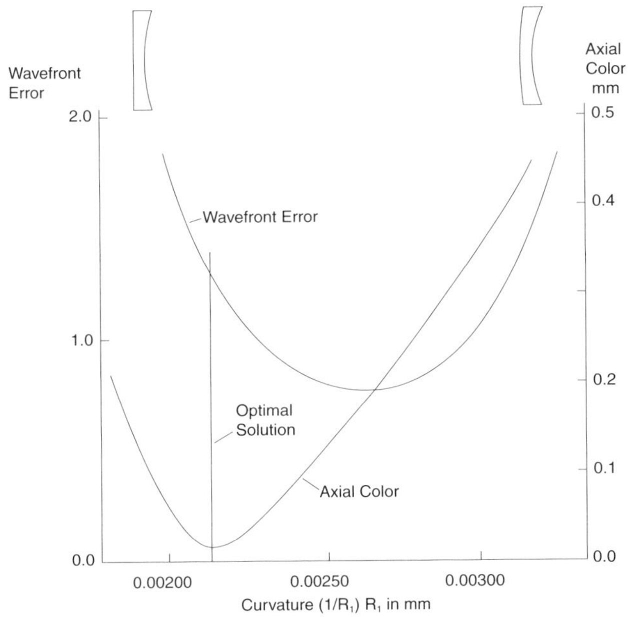

Fig. 8.1 *Analysis for a 14-inch f/12.8 medial with ⅓ diameter type B corrector.*

arrangement of which can be seen in Figure 3.10 in the section on various optical layouts. The graph plots axial color (400–900 nm) and peak-to-valley wavefront residual as a function of corrector bending, expressed as $1/R_1$ (the curvature), where R_1 is in mm, and by definition, the lens power is held constant. The mirror component's radius is also held constant leading to slight focal length changes across the range of bending. The small sketches at the top of the graph show the corresponding corrector shape (starting with a Plano concave) for the extremes analyzed. Light enters the concave surface (the mirror element is not shown). The (peak-to-valley) wavefront error shows the amount of aspheric work (in waves path difference, not surface) required to fully correct this residual. For a given objective aperture, residual axial color and wavefront errors are *f*/number sensitive, thus the graph only represents this specific 14-inch *f*/12.8 system.

Note, for this design, that the all-spherical condition *cannot* be fulfilled. For the final design a bending ratio R_2/R_1 of 10.8 was chosen ($c_1 = 0.00212$). For this ratio the graph shows that axial color is nearly minimized and the residual wavefront error is only 1.4 waves (587nm). This error is best corrected by figuring R_1 of the corrector lens element and it, happily, can be zonal tested at radius. For this design the conic constant is –0.429, an ellipse, thus the figuring method is the same as for partially "parabolizing" an *f*/4.1 surface to a point well under ½ correction. Interestingly, any medial requiring aspheric work will show the best performance in ray trace analysis when the first surface of the corrector is figured.

The second graph (Figure 8.2) shows the Super Schupmann design (see Sec-

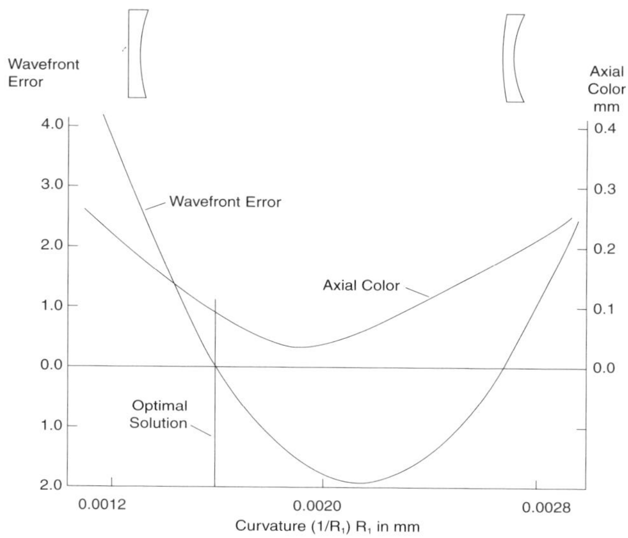

Fig. 8.2 *Analysis for f/10 Super Schupmann with type B corrector.*

tion 5.2) where a Type B corrector is substituted for the normal (Type A) singlet type. Please note, when making comparisons to the previous design, that this design is traced at *f*/10 and the residual wavefront error is scaled differently. Interestingly, at this objective-to-corrector diameter ratio, an all-spherical solution is easily attained; in fact a widely spaced double solution for spherical exists as can be seen in the two zero crossings. This is the same situation that can arise when designing an achromat for zero spherical aberration. Choosing the correct zero crossing is important to the achromat's performance and the same is true here, but for different reasons. If we were to choose the right-side crossing, axial color would double, though remaining well under the Airy disk diameter. However, the chance for ghosting skyrockets as the back surface has just passed through a match with the mirror curvature. Notice that leaving the back surface flat (c_1 = 0.0012674) leaves an acceptable value for axial color of 0.20mm and 3.37 waves P-V residual spherical. Designers of the Swedish 1-meter solar medial found that with quartz elements the corrector back could be left flat as that was so close to optimum for minimizing astigmatism that the manufacturing advantages far outweigh the slight gain from adjusting the bending away from this simplified solution. They figured their objective and corrector separately for spherical for perfect first-focus monochromatic performance, the usual practice for advanced solar medials. Again, the traced wavelengths span the region 400–900 nm.

Builders who may prefer a Type B corrector in the Super Schupmann proportioned design are encouraged to take that path. A prescription for this variation is given in Table 8.1. These curves place the zero spherical crossing on the left in

Table 8.1
Prescription for 10-inch *f*/10 Super Schupmann Variation

Element	Glass	Radius 1	Radius 2	Thickness	Diameter	Air space
1	BK7	56.638 CX	587.61 CX	1.0	10.0 CA	99.40
2	ULE	70.89 CC	Grind flat	0.30	1.0 CA	54.59
3	BK7	24.370 CC	114.49CX	0.472	5.526 CA	0.079
4	ULE	29.035 CC	Grind flat	0.75	5.542 CA	54.575*

* Distance to focus from corrector's first surface. Element # 1=Objective, 2=Field mirror, 3= Corrector lens, 4= Corrector mirror

Figure 8.2. All dimensions are in inches.

It is recommended that the builder stay closer to *f*/12.5 by reducing the apertures or by scaling the design by a factor of 1.25, leaving the apertures as is.

8.5 A Special Case — The Hyper Schupmann

After plotting the curves in Figures 8.1 and 8.2. It became clear to the author that a special case must exist where spherical aberration and color residual are minimized for the *same* bending at some objective-to-corrector diameter ratio. Intuitively the ½-diameter corrector came to mind as it mimics the full-size corrector in that the refraction angle per pass is the same as the objective's refraction angle, thus as with the full size corrector, a high order balance is achieved. A ray trace of a 10-inch *f*/10 with a half-diameter corrector confirmed that the color shift, after minimizing at a particular bending, is nearly as mathematically perfect as for a full-size system (see Chapter 5). It reached a minimum of about 0.01 mm over the spectral region 400–900 nm. Some aspheric correction was necessary, but amounted to only 0.6 wave optical path difference (OPD). Now the search was on. By extrapolation, the data indicated that a diameter ratio (defined as EFL/BFL) of 2.117 or 47.24% would come close to a simultaneous solution for spherical and color. The corrector's lens element was bent for zero spherical and the color and wavefront error was then explored using small bending steps on either side of the all-spherical solution. After plotting the data (see Figure 8.3) it was clear that the best possible mathematical color correction seen at 50% diameter was slightly degraded, it now showing a value of about 0.015 mm, still unbelievably tiny. Note that now the minimum color was very close to the zero crossing curve for spherical, and almost no mathematical gain is made by perfect coincidence. (A good optimization project for the student.) In fact, as one moves closer to perfect coincidence the corrector gets smaller and, as with proportionally smaller correctors, the color error slowly looms up.

In the author's opinion, the solution obtained is a reasonable compromise and provides the ideal corrector for the 10- to 20-inch instrument. The writer has named this all-spherical design variation the Hyper Schupmann in keeping with its nearly mathematically perfect axial correction. A scale corrector drawing is included on the graph in Figure 8.3. The single line to the left of the lens is the con-

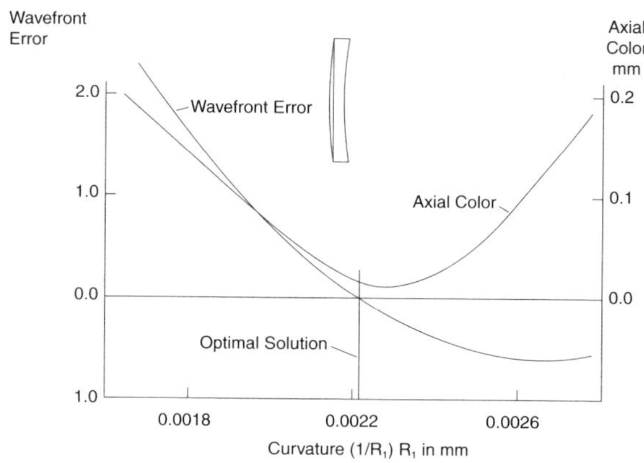

Fig. 8.3 *Analysis for 10-inch f/10 Hyper Schupmann.*

Table 8.2
Prescription for 10-inch *f*/10 Hyper Schupmann

Element	Glass	Radius 1	Radius 2	Thickness	Diameter	Air space
1	BK7	56.638 CX	587.60 CX	1.0	10.0 CA	99.400
2	ULE	64.449 CC	Grind flat	0.3	1.0 CA	47.244
3	BK7	17.809 CC	76.870 CX	0.472	4.80 CA	0.118
4	ULE	23.543 CC	Grind flat	0.60	4.80 CA	47.244*

* Distance to focus from first surface of corrector.

cave reflecting surface.

The Hyper Schupmann is ideal for the serious observer working over a small scientific CCD field, thus great care should be taken in the mechanical construction especially in the area surrounding the focal region to minimize the corrector and objective tilts to just meet these aims. A corrector tilt of no more than 0.6 degree is recommended. The compensating orthogonal objective tilt will be similar and as with any medial is fine-tuned on the stars.

Table 8.2 gives a baseline prescription for a 10-inch *f*/10 Hyper Schupmann, allowing the ATM to build as-is or scale the values to various other apertures and *f*/numbers. The author makes a plea to all builders: if possible, work closer to *f*/12.5 by scaling appropriately.

Who will be the first to try a Hyper Schupmann?

Chapter 9
Schupmann's Brachymedial

9.1 Introduction

Although the greater part of this book is devoted to the medial design, to leave out the brachymedial (i.e., shortened medial) story would ignore a design not only of great fascination, but also of potential scientific usefulness. The author's comments in his1984 book on the Schupmann telescope suggested little hope for the brachy idea. As we shall see, the faint embers of the brachy have flamed-up!

This design is very closely related to Fr. William Hamilton's invention for color-correcting a single-element crown objective. Hamilton was granted English patent, #3781 dated April 12, 1814, titled "Certain Improvements in Optical Instruments or Apparatus." It is not clear if Schupmann knew of Hamilton's work as he does not seem to be mentioned anywhere in his 1899 book *Die Medial-Fernrohre,* and yet Alexander Rogers' work regarding the development of the all-lens dialyt is![1] The Hamiltonian refractor, as it is now known, originally utilized a flint corrector backed by a speculum mirror, which returned the cone of rays to focus just short of the objective where they were reflected by a diagonal mirror as in the Newtonian arrangement. Hamilton, like Schupmann, wanted to color-correct a large, single-element objective; but their respective designs were driven by different problems. In the case of Hamilton, it was the unavailability in England at that time, of large striae-free flint blanks to make a standard refractor; with Schupmann it was the quest for a perfectly color-corrected telescope not obtainable with a standard doublet. Hamilton's design succeeds (compared to a crown-flint doublet) on-axis but the gigantic lateral color forced the design into obscurity. In any event, Schupmann's exploration of this form of refractor resulted in some improvements as well as suggestions for correcting lateral color. It is also interesting to note, in connection with Hamilton, that Schupmann patented both the brachymedial and the medial both in Germany and America and these patents were idea-centered rather than prescription-dominated.

The brachymedial for some reason attracted the attention of those who read Schupmann's 1899 US patent even though it and the medial were equally well shown and described. One reason for this may be the apparent simplicity of the brachy design. Therefore, early published articles[2,3] in America centered on this

[1] Rogers, Alexander, "On the Construction of Large Achromatic Telescopes," published in Vol. 3 of *Memoirs of the Astronomical Society of London*, 1829, pp. 229–233.

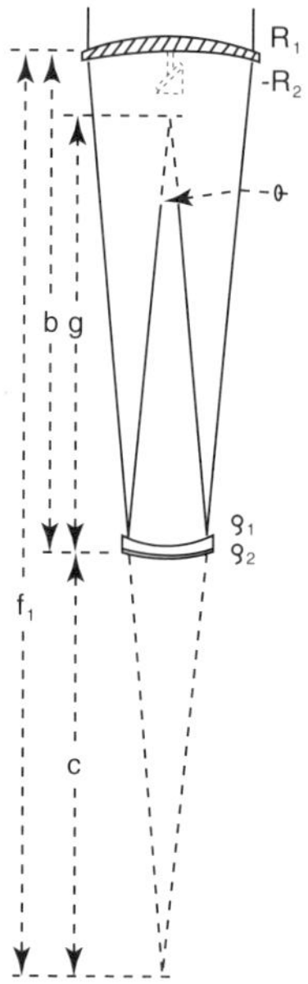

Fig. 9.1 *Schupmann's "Type A" corrector. From "Die Medial-Fernrohre."*

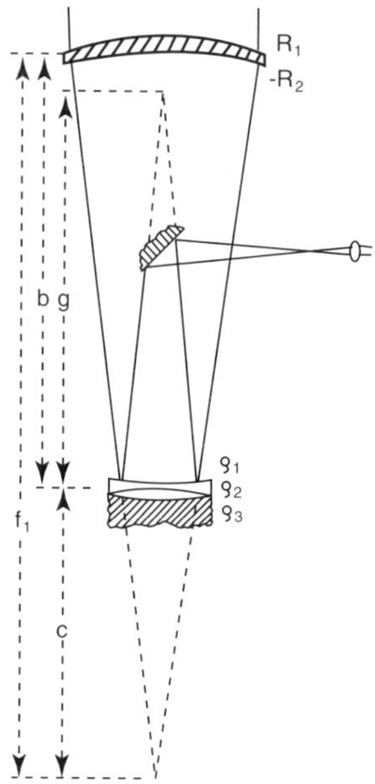

Fig. 9.2 *Schupmann's "Type B" corrector. From "Die Medial-Fernrohre."*

form. It is a curious design as it is centrally-obstructed, thus taking away one of the strong points of refractors in general. As with any telescope design, it is possible to devise an obstruction-free eccentric pupil version, although construction and alignment may prove troublesome.

9.2 The Optical Design Layout

Delightfully simple in appearance, the brachymedial optical train consists of a single-element positive meniscus objective lens of crown glass with its convex surface skyward followed by, in the simplest arrangement, a meniscus corrector lens located at the bottom of the tube. The latter's concave face is skyward, and its rear, convex surface is silvered to reflect the cone of light back through and out the corrector where it continues to converge towards a focus some distance up the tube. A diagonal mirror intercepts the cone and directs it to focus outside of the tube. Schupmann called this version "Type A". His "Type B" layout employs a separated concave mirror to double-pass the light through the corrector lens. Schupmann's diagrams in Figures 9.1 and 9.2 show the basic layout for both types. By shaping the curves correctly, the axial color aberration of the simple objective is

[2] Hellman, Carl A., "Gleanings for A.T.,M.s" edited by Earle B. Brown, "The Schupmann Telescope," *The Sky*, September 1941, pp.14-15.

[3] Brown, E.B., "Gleanings for A.T.M.s" edited by Earle B. Brown, "Report on a Schupmann," *Sky and Telescope*, April, 1942, p. 20.

nearly balanced by the corrector, with color of the opposite sign. Although employed more succinctly in the medial, this method of color correction is the fundamental principle underling all variations of Schupmann-type telescopes. Schupmann's invention and all its many iterations, are known throughout the optical design community as the "Schupmann Principle". It is only when the brachy is examined in detail that its outward simplicity vanishes—in fact, true perfection of this design concept has only recently been achieved.

9.3 Good Points and Drawbacks

Unlike the standard medial, the brachymedial employs no tilted elements. Schupmann probably worked up this design before the medial as a result of his attempts to improve the dialyt—after all, the corrector is positioned as would be the second lens set in a dialyt. On the face of it the original brachy seems like the ideal planetary instrument with its reasonably short tube and nice long focal length. The correction of spherical aberration can be accomplished with spherical surfaces but not without harm to field performance. Curvature of field is fairly shallow and, unlike the medial, is inward curving. The corrector can be multi-element, such as Type B, for greater aberration control, but light loss begins to add up in such arrangements making antireflection coatings mandatory. Schupmann showed that, unlike the normal achromatic refractor with its crown and flint objective and huge secondary color, his medial and brachymedial can utilize the same glass combinations yet provide a high degree of color correction. His use of a flint corrector allowed the curvatures to be greatly reduced giving better control of various aberrations. Because the lens powers are so low compared to a normal refractor, the brachy's residual color (C to F) is typically ⅕ that of the traditional refractor of the same aperture and focal length. This is well within the diffraction disk at the typical focal ratios employed in the brachy designs of old. As the ratio of objective-to-corrector diameter is reduced, as in modern interpretations, the usefulness of various light flints for the corrector disappears, both in the brachy and the medial form.

Early design versions of the brachy required a spider to support the diagonal mirror which for practical reasons, in view of the giant brachy's then proposed, positioned this mirror about mid-tube. This allowed the beam to exit the tube down the declination axis. A giant brachymedial presently in use in Rathenow Germany is of this form, and is discussed in Section 9.9. Optical layouts such as this further increase the diagonal blockage, as it is located not at the aperture but approximately ¼ of the way to focus. However, the harm from this should not be exaggerated as the mirror for the inherently narrow field can still be quite small. Schupmann recognized the optical drawback of a mid-tube viewing position and, in his book *Die Medial-Fernroher*, showed schemes with the diagonal (prism) supported by a stalk attached to the objective lens. This would work very well with small amateur instruments and finds full implementation in Honders' brachymedial designs described later in this chapter.

Unlike the medial, the brachy is without an intermediate focus and associated re-imaging optics. This leads to significant optical decenter or "beam walk" on the compensator element for off-axis bundles due to its great distance from the objective. This introduces devastating lateral color in the images, which limits the field of fine image quality to a size that just includes the diameter of Jupiter, requiring careful centering indeed! This aberration, along with residual coma, is correctable with a multi-element, nearly zero-power, lens located a short way inside focus. The introduction of this field lens limits the axial color correction to about ⅕ that of an equivalent Fraunhofer objective. Schupmann devised "Kompensating Okulars" for these oblique aberrations, not a very satisfactory solution in modern times. The only way to reduce beam walk is to reduce the distance to the corrector. It is possible, with modern interpretations, to correct the lateral color completely with only a weak single-element positive field lens.

Spherical aberration is easily corrected using all-spherical surfaces with the objective usually taking the form of a positive meniscus, as in Schupmann's original designs. As mentioned earlier, the corrector or "Kompensator" can be a single-element component, or take the form of a double-pass "Littrow" arrangement. The dividing of lens powers, in this case, favorably impacts the level of correction obtainable with spherical surfaces.

The objective lens shape factor for zero coma is at odds with best spherical correction, so some aspheric work may be needed for coma-free versions. Astigmatism remains uncorrected for early designs and is similar in strength to a normal refractor of like focal ratio, and optically harmless in the brachy.

Schupmann's traditional brachymedial provides the amateur ray tracer a wonderful system to explore and, as you will see, the brachy story has just begun!

9.4 Old-Style Brachy Designs for Amateur Construction

The author has worked up three brachymedial design prescriptions for amateur construction. The first two far out-perform the standard Fraunhofer-type refractor of the same diameter and focal ratio, and will compete favorably, visually, and over a narrow field, with modern apochromatic instruments at ⅒ their cost. The third design outperforms the first two by using an unusual glass for the corrector. These designs can be hopped-up with optimization programs and are presented here as practical introductory instruments. These designs cannot compete with the standard medial because their axial color over the spectral range of a typical CCD vastly exceeds the diffraction disk. Furthermore, a special cemented field lens must be made for satisfactory performance.

9.4.1 Type A Brachy Design

The first example is a Type A all-spherical, 6.8-inch *f*/23.5, instrument. The correction over a 0.3-degree full field is quite good and with careful work will make a nice planetary scope. The visually important C and F lines are corrected to 0.02 waves peak-to-peak and are focused co-planar. The d line falls slightly long re-

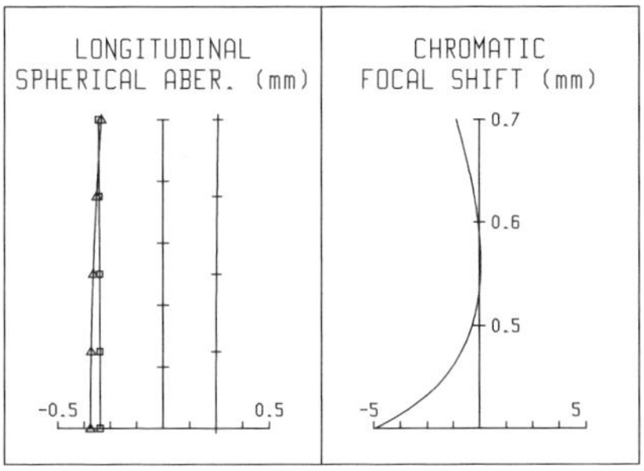

Fig. 9.3 *6.8-inch, f/23.5 Type A brachy.*

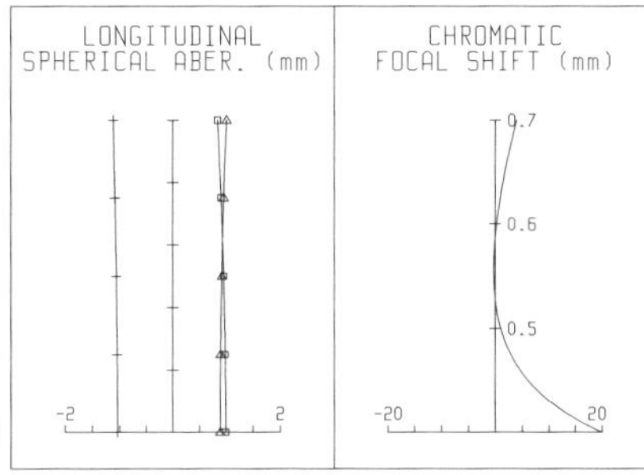

Fig. 9.4 *For comparison purposes, the same data for a 6.8-inch, f/23.5 Fraunhofer doublet.*

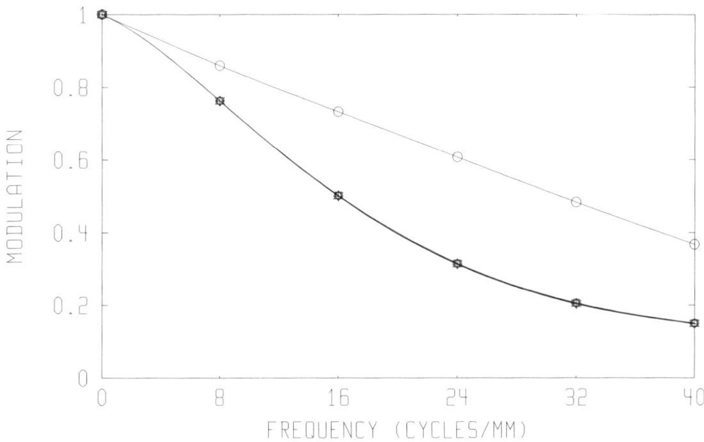

Fig. 9.5 *MTF for 6.8-inch, f/23.5 Fraunhofer doublet.*

quiring that a CCD be focused out about ½ mm for the green for RGB color imaging. In practice a compromise focus position just 0.3 mm outside of C and F shows a near theoretical modulation transfer function (MTF, see Section 4.1) and will answer nicely for all work including double star measures. Figure 9.3 shows the longitudinal spherical plotted against ray height for the C-squares (656 nm), d-crosses (587 nm) and F-triangles (486 nm) lines; and color curve (chromatic focal shift) plotted against the wavelength in microns. Note that in Figure 9.3 (left panel) that the colors are coded as follows: 656 nm = squares, 587 nm = crosses, 486 nm = triangles. This convention is maintained throughout this chapter. The same data for an equivalent Fraunhofer doublet are shown in Figure 9.4 for comparison. Take care to note the horizontal scales! Figure 9.8 is an MTF curve for the Fraunhofer doublet and is also instructive when comparing various Brachy designs.

Table 9.1 gives the prescription for the Type A instrument and Figure 9.7 shows its MTF. Figure 9.7 shows a perspective view of its layout.

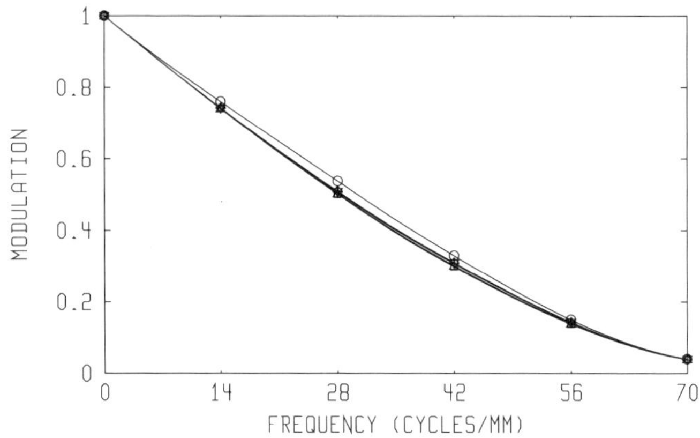

Fig. 9.6 *MTF curve for Type A brachymedial detailed in Table 9.1. Field = 0.2 degree, Wavelengths C, d, and F. Upper curve with small circles represents a theoretically perfect system.*

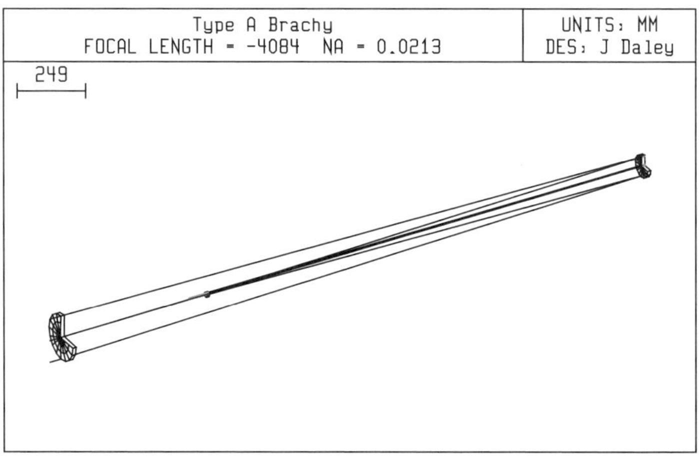

Fig. 9.7 *MTF curve for Type A brachymedial detailed in Table 9.1. Field = 0.2 degree, Wavelengths C, d, and F. Upper curve with small circles represents a theoretically perfect system.*

When using a CCD for white-light imaging of the planets it may be necessary to use an IR blocker and a very light yellow filter, because sharp imaging in this design begins to fail at the ends of a typical CCDs spectral coverage.

An objective of BK7 and single element corrector of F2 (flint) is employed in this design. Although the use of flint leaves some residual axial color, the design is almost entirely free of spherochromatism. Silver is the coating of choice for the reflecting surface of the corrector and if the first surface is anti-reflection coated the light loss in this element is less than 5%. In this design there are two concave surfaces to help with radii and figure control. The 13.5-foot focal length will just fit in a 8-foot plywood tube with the cells bolted on the outside ends, still a manageable system. For best performance the diagonal mirror (not shown) should be

Table 9.1
Prescription for 6.8-inch *f*/23.5 Type A Brachymedial
Corrected for lateral color.

Element	Glass	Radius 1	Radius 2	Thickness	Diameter	Air space
1	BK7	73.248 CX	430.75 CC	0.625	6.85	94.21
2 R_2-silvered	F2	43.779 CC	113.787CX	0.512	3.379	70.67
3	F2	Plano	1.929 CX	0.125	0.700	Cement
4	SK2	1.929 CC	1.929 CC	0.125	0.700	Cement
5	F2	1.929 CX	Plano	0.125	0.700	2.768-focus

All dimensions are in inches. CC=concave, CX=convex, Airspace =distance to next surface: vertex-to-vertex or vertex-to-focus.

no larger than 20mm on the minor axis, which will give full illumination over a 13.4 mm field diameter. Obstruction by the diagonal is only 16% of the aperture diameter, or 1.8% of its area, so the performance will be but little degraded. Making a single spider vane of 0.020"-sheet steel, which is circularly curved through a 90° arc within the beam, will render diffraction effects invisible.

A cemented 3-element field group is located before focus giving a BFL of 72.17 mm. This lens set corrects lateral color and most of the, otherwise, harmful coma. The instrument will not be satisfactory without this field group, which must be in place for all figuring involving autocollimation. Glass types for the field group are F2 for the outer elements and SK2 for the internal lens. All internal radii are the same and the outer surfaces are flat so they may be ground on each other. No further magnification is required for CCD planetary imaging as the focal ratio satisfies the Nyquist sampling criterion for pixels in the 10-micron range. This design even shows marginally acceptable performance at 8-inch aperture with the same curves. Of course the other lenses will need enlarging by the same ratio.

9.4.2 Type B Brachy Design

The second design is a Type B all-spherical 6.8-inch operating at *f*/27.9. The useful field is about half that of the first design but the critical planetary region shows geometric spot sizes about half as large—essentially perfect for the amateur size CCD. The spectral region of sharp imaging is somewhat extended over that of the Type A (see color curves in Figure 9.8) making CCD imaging more practical. For visual use the color correction leaves little to be desired. However, spherochromatism (and coma for larger fields) does limit the potential of this design. Again the focal ratio satisfies the Nyquist sampling frequency for direct CCD planetary imaging. As with the first design, the aperture can be pushed to 8-inch leaving the curves as is.

This version is especially interesting to build as the radii for corrector and separate reflecting mirror are all concave and equal, thus making manufacturing and knife-edge null testing at the radius of curvature of each surface easy. The back of the objective is also concave and again a near perfect sphere is possible with standard knife-edge testing. This test of the objective requires an area, includ-

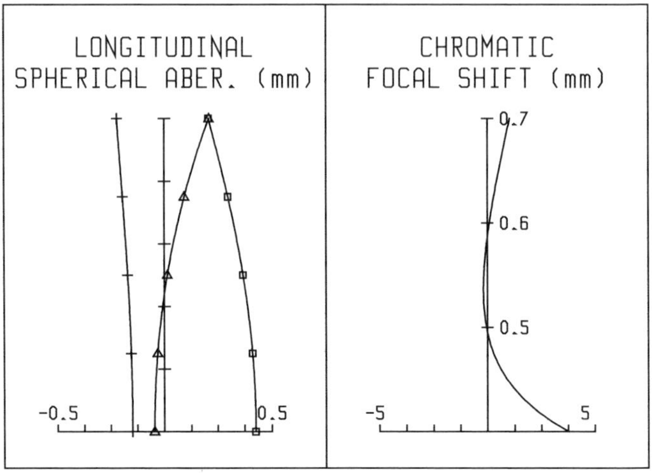

Fig. 9.8 *6.8-inch, f/27.9 Type B brachy.*

ing seating room, no longer than 15 feet. Use of radius rods, as with the corrector elements, will keep the curve exact. The objective's first surface is figured by autocollimation of the *entire* system or, alternatively, the grinding tool could be made a spherical test plate. As in the medial both lenses are BK7, but BSC2 can also be used with little or no design change. The reflecting mirror should be of low-expansion glass, so Zerodur or quartz is recommended. The curves are designed for the mirror and lens to be in contact, rim to rim, against lightly ground flat lands. This will protect the reflecting coating from atmospheric degradation and dust. A good anti-reflection coating on the corrector lens and a special enhanced coating on the mirror will really boost the light transmission. For those with the talent for this project there should be no need for fine construction details that are not otherwise covered here or in other chapters but do watch to keep the wedge under strict control on both the corrector and objective lens.

This design is also corrected for lateral color with a field group yielding a BFL of 78.23 mm. This lens is a 2-element design with matching internal radii. The outer surfaces have separate but weak curves. Cementing the elements will increase the transmission significantly. The glasses are F2 and SK2 as in the first design. Due to the variation of magnification with color (lateral color), planetary work *without* a correcting field group is, as a practical matter, hopeless. (The image of Jupiter would need to be accurately centered and tracked to about ¾ of the gas giant's diameter).

The diagonal mirror can, and should, be very small; around ⅝-inch (1⁄14 the aperture diameter) will cover the useful field easily. Since the diagonal is located some 700 mm behind the objective it is tempting to think in terms of supporting it with a lightweight stalk from the objective. This stalk should be a full ⅝-inch diameter and made from a specially turned thin-wall aluminum tube with a turned plug and bolt set-up fitting a (suggested) ¼-inch hole drilled in the exact center of

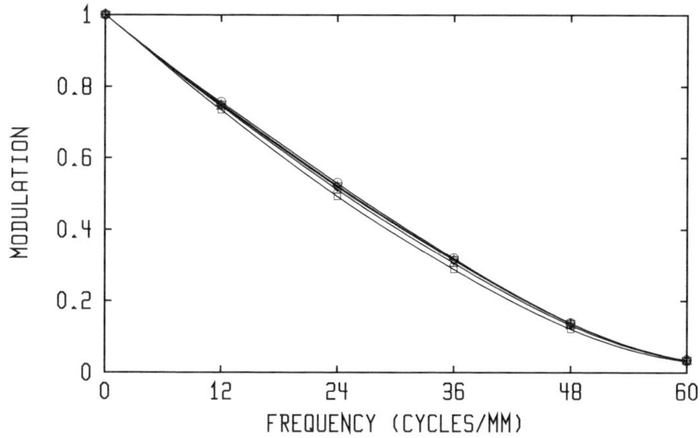

Fig. 9.9 *MTF curve for Type B brachymedial in the prescription detailed in Table 9.2. Field 0.1 degree, Wavelengths C, d, and F, nearly perfect over limited-field visual use.*

the objective. The ⅝-inch diagonal mirror and adjustments need to be as light as possible, and the whole set-up will demand the best instrument making abilities and a full understanding of how to dampen any potential resonance of such a structure. Alternatively, a spider as in the first design is just fine and easier on the nerves!

This telescope can be built in just a few months by any first-class optician and would be great fun to experiment with, such as computer optimizing the little field group lenses to get a bigger field and the like. Other experiments could include design variations, which use flint glass for the corrector. Optical "bending" of the corrector lens (necessarily along with the objective) can be tried to improve the residual axial and field aberrations but two more grinding tools will be needed for construction. The amateur with ray trace capabilities will delight in the fun this design presents.

As with the first design, the optical tube is made from standard 8-foot,¼-inch mahogany ply. Building a scaled-down 4-inch version of the Type B design would make a nice project for those of an experimental nature. Both designs work best at the focal ratios specified, for larger instruments the critical builder should scale-up all the values in the prescriptions.

Table 9.2 gives the prescription and Figure 9.9 the MTF curve. A perspective drawing is shown in Figure 9.10.

9.4.3 A Special Type A Brachy Design

The third brachymedial design is a special Type A. This design came about after a discussion with Roger Ceragioli who has an abiding interest in unusual refracting telescopes. He suggested the use of K10 for the meniscus corrector. It only required a few moments' ray tracing to see the huge improvement this glass made. In the example detailed here we see color correction of a quality that nearly spans

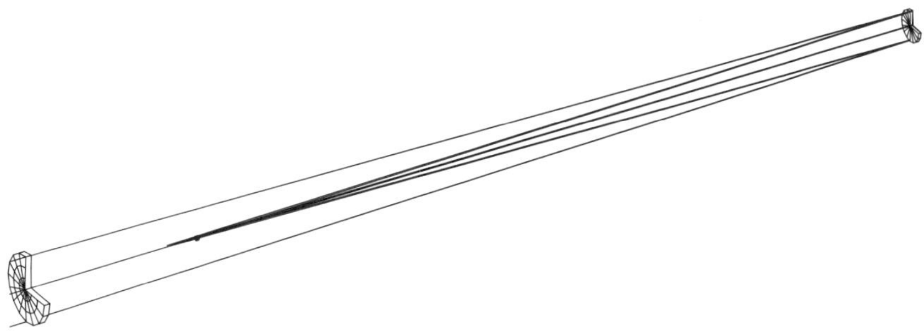

Fig. 9.10 *Isometric Type B brachymedial.*

Table 9.2 Prescription for 6.8-inch *f*/27.9 Type B Brachymedial

Element	Glass	Radius 1	Radius 2	Thickness	Diameter	Air space
1	BK7	52.634 CX	130.669 CC	0.666	6.85	94.882
2	BK7	64.913 CC	64.913 CC	0.236	3.165	0.236
3	ULE	64.913 CC	Mirror	~0.5	3.165	76.772
4	F2	27.559 CX	0.905 CX	0.125	~0.750	Cement
5	SK2	0.905 CC	11.811 CC	0.125	~0.750	3.071-focus

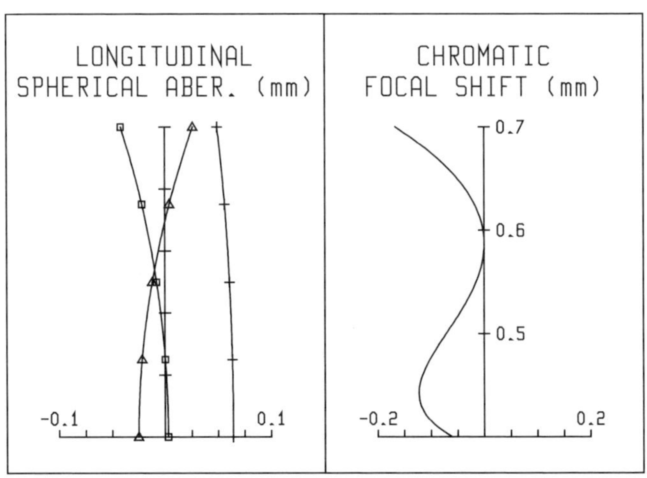

Fig. 9.11 *Type A brachy employing K10 glass for the corrector.*

the spectral range of a CCD. See Figures 9.11 and 9.12 to compare with the previous design examples.

This design is a serious amateur instrument and is satisfactory with the curves shown up to 14-inch aperture! At 12-inch aperture the on-axis performance is impeccable. The maximum field of superb definition is limited to 1⁄10 degree (~8 Jupiter diameters), which overfills most affordable CCDs. This design requires

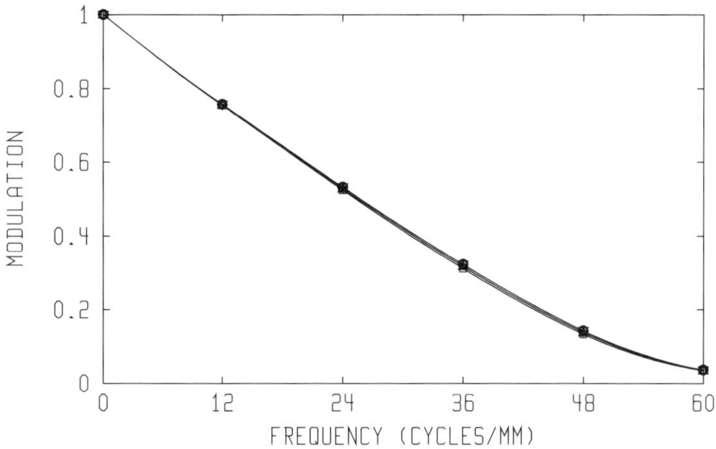

Fig. 9.12 *MTF curve for special Type A as detailed in Table 9.3. Field dia. = 0.1 degree, wavelengths C, d, and F.*

some light aspheric work, easy for any advanced amateur. For the ray trace study the objective's first surface was aspherized. Again the lateral color is corrected with a 3-element group similar to the previous Type A design. This lateral color correction is quite good between C and F, but begins to gracefully fail, as a greater span is demanded. One major difference from the two previous design examples is the focus position which, if not intercepted with a diagonal mirror, lies outside the objective. Now, by attaching the diagonal on a short stalk cemented to the back of the objective lens, the converging beam is focused neatly well outside the tube. Although this puts the focus much higher and will require a tall step-up, the diagonal mirror can be made very small. This focus position puts the observer at the objective, so harmful thermal currents can influence the image quality. This objection equally applies to a cooled CCD as its dissipated heat load can find its way into the light path. The author has tried to drive the focus down the tube for a longer diagonal stalk, but the correction begins to fail fairly fast. For scientific use one should consider a round diagonal mirror mounted a on a long stalk. The mirror would then be slightly angled to send the cone well down and out the tube wall emerging at, say, 30° to the tube surface where a robust, lead-screw driven, focus stage supporting the CCD would be located. This would place the heat source well away from the aperture and make object centering much easier. Perhaps other designers can bring further improvements to this concept. OSLO MTF data for a manageable 9.1-inch *f*/27.2 version are shown in Figure 9.14. Table 9.3 gives the design prescription and a perspective drawing is included in Figure 9.13.

Many lens adjustments will be required for these designs and, ideally, should be remotely performed while looking in the eyepiece. They include tilt and focus adjustment of the corrector to both optically center the system and to tune-out axial color. Laser alignment tools are almost a must for the brachy. The field groups in Type A, Type B, and the special Type A must be provided with adjustment

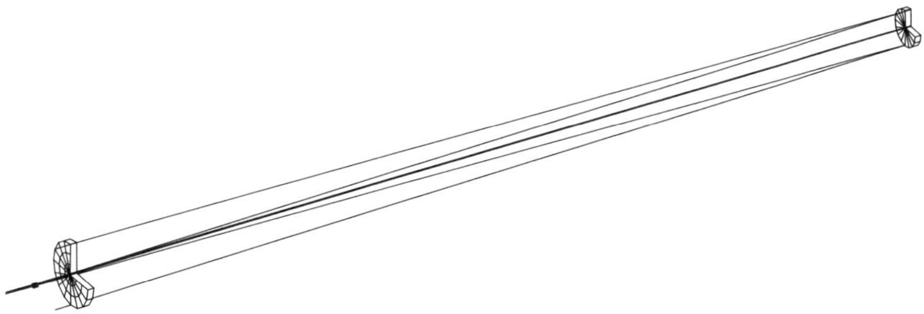

Fig. 9.13 *Isometric of special Type A brachymedial. Note that the focus now extends beyond the objective. See text for suggested diagonal mirror arrangements.*

Table 9.3
Prescription for 9.1-inch *f*/27.2 Special Type A Brachymedial

Element	Glass	Radius 1	Radius 2	Thickness	Diameter CA	Air space
1	BK7	104.346CX	367.259 CC	1.000	9.00	124.778
2	K10	62.067 CC	176.539CX*	0.838	5.190	130.791
3	F2	Plano	3.740 CX	0.193	0.582**	Cement
4	SK2	3.740 CC	3.740 CC	0.193	0.582	Cement
5	F2	3.740CX	62.3 CC	0.193	0.582	4.062-focus

*Silvered surface.

**Minimum dia. for 0.1 full field dia. examined. Recommend this cemented lens group be 1.0-inch diameter for all 3 designs presented. This design requires a weak (conic constant = 0.23) aspheric figure on R1 of the objective.

along the axis as well as laterally to null lateral color. These adjustments, fortunately, are readily at hand. Other than the tilting of the corrector, which must be remotely adjustable for nightly tweaking for optical centering, most adjustments, once completed, should be more or less permanent. The author has no experience building a brachy; however, there seems to be no especially difficult part other than the mechanics of alignment. The first two designs have their eyepieces located just a tad above eye level when observing overhead, assuming the corrector cell just clears the observatory floor. This requires an approximately 6- to 12-inch step-up platform for most adults.

9.5 The Honders Variation — A Powerful New Interpretation

Just as the author was beginning this section, word came that Klaas Honders of the Netherlands had passed away. The author got to know Klaas very well even though our only contact was via the Internet. He was very enthusiastic about an up-to-date book on the Schupmann and was excited about the prospect of including his newest ideas and designs. His designs published here are a tribute to his optical skills and amateur spirit.

Table 9.4
Prescription for Honders 8-inch *f*/7.38 Planetary Brachymedial

Element	Glass	Radius 1	Radius 2	Thickness	Diameter	Air space
1	BK7	216.535CX	529.09 CX	0.630	8.0	53.150
2	BK7	70.614 CC	112.181* CX	0.630	7.531	54.134
3	BK7	10.973 CX	141.421 CC	0.287	1.86**	4.297

*Silvered surface.
**CA based on 1 degree full field diameter

Fig. 9.14 *Klaas Honders, 1951–2003. Courtesy Klaas Honders.*

Klaas became interested in the possibilities of Schupmann's brachymedial and propelled the design to a level of perfection never dreamed of. It is interesting to note a US parallel to Honders' ideas: namely the freeing-up of the relative diameter of the corrector for the smaller amateur instruments. As a background historical note: Schupmann was interested in building huge, 1.5-meter aperture, medial refractors and naturally concentrated his efforts on the design of relatively small correctors. His designs usually employed multi-element correctors of around ⅛ the objective diameter and this ratio was maintained even in all but one of the small German instruments to follow despite obvious shortcomings. Exploration of the design in the US by Edwin Olson led to the remarkable full-aperture corrector medial system and is described in Sections 5.5 and 12.8. Schupmann never explored a brachy corrector diameter beyond about ½ the objective diameter. As with Olson's medial, Honders designed brachy systems with near full-aperture correctors. His designs are, in fact, at the outer limit for the term "brachymedial" as the instrument length is only slightly "shortened." The focal ratio of the "objective" is highly altered by the corrector; in fact the corrector provides the majority of the power in the system. This leads to the question: has the objective become the corrector? His designs still qualify for the term "brachymedial", however, the important thing to note is that the design follows the Schupmann principle of color correction and thus is a true Schupmann telescope.

9.6 The Honders Planetary Brachy

Table 9.4 gives the prescription of the Honders all-spherical planetary brachy. The ray trace results are shown in Figure 9.15 and a layout drawing in Figure 9.17. The overall optical perfection of this 8-inch system leaves little to be desired. The modulation transfer function (MTF) shown in Figure 9.16 is nearly indiscernible from what is theoretically possible. Lateral color is insignificant with an edge of the field value of 0.0001 mm. It has an absolutely flat focal surface (Petzval radius = 2.7 meters!), with image quality diffraction-limited everywhere over a huge 1°

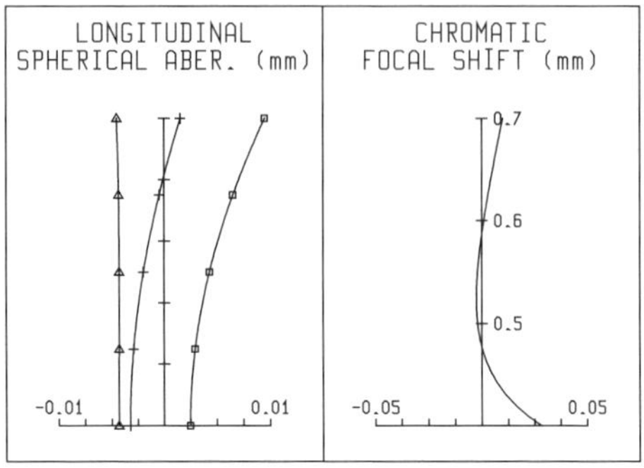

Fig. 9.15 *Longitudinal spherical and color curve for the Honders planetary brachymedial. Horizontal scales in mm. Vertical divisions (left) are fractions of aperture radius and (right) wavelengths in microns.*

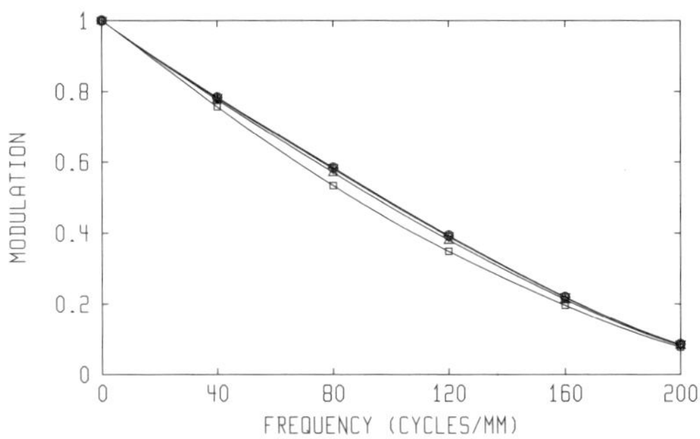

Fig. 9.16 *MTF curve for the Honders planetary brachymedial for 1 degree full-field diameter.*

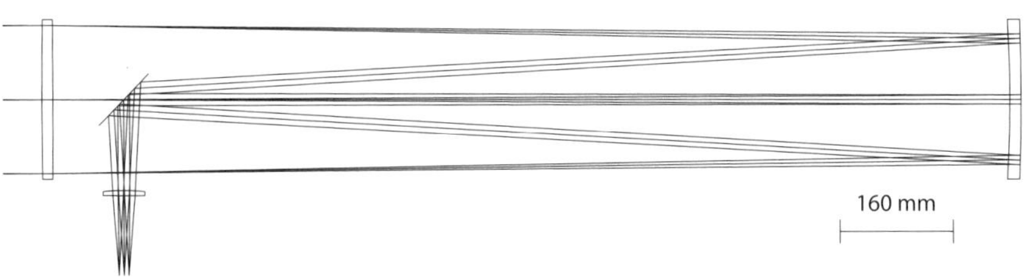

Fig. 9.17 *Layout of the Honders planetary brachymedial.*

full-field diameter. At best planetary field focus the correction is 0.004 wave, peak-to-valley.

This comparatively fast *f*/7.38 system is astrometrically perfect as the distortion is only 0.02%, and thus would provide ultra-precise asteroid position measurements when coupled to a large-format CCD. Planetary work would require minimizing the diagonal mirror size, and since there need be no spider the performance will be impeccable. The diagonal mirror position (not shown) should be chosen such that the field lens just clears the slightly converging beam from the objective. This, of course, minimizes the diagonal's obstruction. CCD imaging of

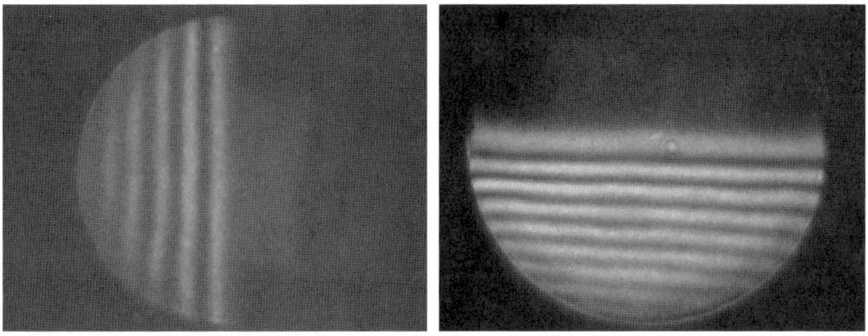

Fig. 9.18*Null test interferogram of Honders' 8-inch corrector showing fringes positioned vertically and horizontally.*

Table 9.5
Prescription for the Honders 6-inch *f*/2.7 Brachymedial Camera

Element	Glass	Radius 1	Radius 2	Thickness	Diameter CA	Air space
1	BK7	61.295CX	115.157 CX	0.472	6.0	13.680
2	BK7	18.727CC	30.797* CX	0.472	5.68	13.937
3	BK7	4.652CX	118.751 CC	0.236	1.705**	1.707

* Silvered surface.
**CA based on 3 degree full-field diameter.

the planets will require a top quality 3x Barlow pushed to 3.5x for 10-micron square pixels.

9.7 The Honders Brachy Camera

Nothing in the author's experience with the Schupmann principle could prepare him for this development. Here, in one step, all other astrographic designs are swept aside. This 6-inch *f*/2.7 system (see Figures 9.19 and 9.20) has all the features any wide-field CCD astro-imager could ever desire. The field is dead flat and nearly distortion free. Color correction is near perfect over a huge wavelength span. Lateral color is reduced to an incredibly low value of only 0.0005mm at the field edge! Ghost images have been examined and found to be, for all practical purposes, undetectable. For serious research we find the design perfectly suited to asteroid search programs and variable star work where the wide CCD field will almost always contain a comparison star. CCD work will require a Newtonian arrangement; however for film work a special film holder could be placed outside of a centrally-cored objective, which holds the final lens. This would greatly minimize the obstruction. For astrophotography, the author believes no finer all-spherical design exists! Table 9.5 gives the prescription and Figure 9.19 shows the spherical and color curve. The MTF curve shown in Figure 9.20 is remarkable for a system this fast. Scaling the design to an 8- or 10-inch aperture is within the capabilities of advanced amateurs. Working closer to *f*/3 would be a wise idea at the larger apertures though. Each proposed camera should be ray traced to find the

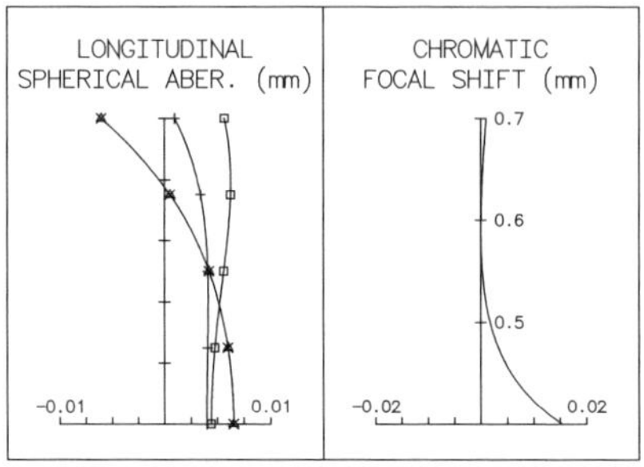

Fig. 9.19 *Longitudinal spherical and color curve for the Honders brachy camera. Traced wavelengths are 656nm (squares) 587nm (crosses) 486nm (triangles).*

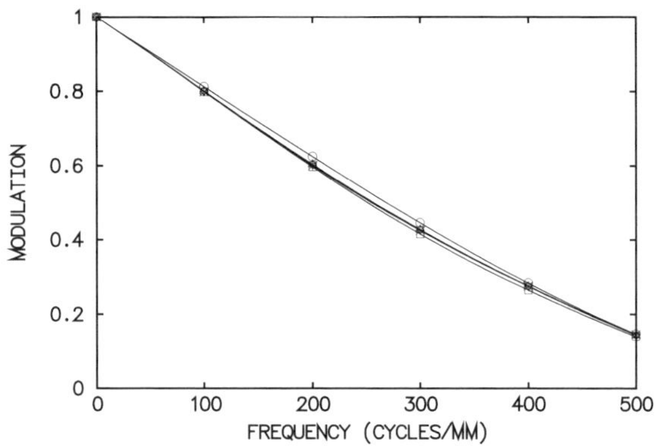

Fig. 9.20 *MTF curve for the Honders brachy camera (3 degree full f/Field) showing near theoretically perfect performance.*

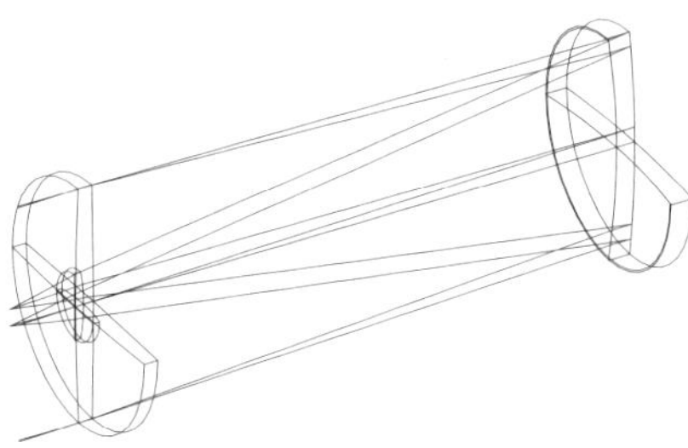

Fig. 9.21 *Wire frame view of the Honders 6-inch f/2.7 brachy camera. See text for use without a diagonal mirror.*

best compromise to meet the user's objectives.

This section was to be a generous contribution by Klaas Honders and a reflection of his experiences building his planetary version briefly described in the last section. Klaas had made great progress on his planetary instrument with the objective complete and the corrector in final figuring. Interferograms of the corrector (see Figure 9.18) showed its surface accuracy to be better than 1/10 wave in red light, and his last communication with the author mentioned silvering the re-

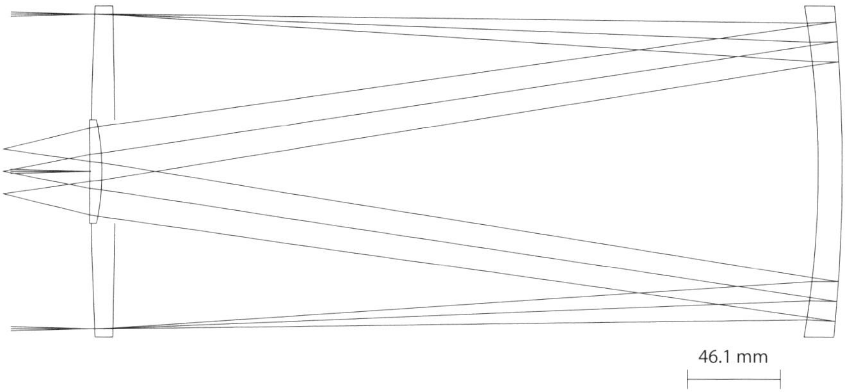

Fig. 9.22 *The Honders 6-inch f/2.7 brachymedial camera configured for film use over a 3 degree FOV. The field lens is best mounted in a threaded barrel running in a threaded cell which provides fine focus. This cell is retained in an accurately bored hole in the objective. A ~1-inch film chip held flat in a subsidiary dark-slide chamber would be clipped to the cell. The central obstruction blocks 10.5% of the incoming light. It may be possible to mount a small (futuristic?) CCD in place of the film.*

flecting side of the this element. The author has no direct experience with the special problems that may be encountered and methods to overcome the various difficulties in this specific design. The corrector is, of course, very sensitive to thermal effects, but not impossibly so, as Honders demonstrated. We agreed that adjustments must be provided in the finished instrument to center the objective and tilt and translate the corrector, and also some similar adjustment for the field lens and diagonal mirror. As with any medial design, anyone tackling this project must be a top-notch optical type character. It should also be clear to the reader that this design does not lend itself to cardboard tube construction, so the builder had better be one good machinist! Keep in mind that the recommended adjustments are necessary for both the planetary and astrographic versions. Other optical work common to any Schupmann telescope is covered in the section on grinding, polishing and testing of the Standard Medial Design.

9.8 Schupmann's Brachymedial Ideas

A historical find by Uli Mayer gives us a rare view of Schupmann's idea for a possible commercial version of a low-power brachymedial. Detailed drawings of the mount, tube, lens assemblies and letters, all in Schupmann's hand, were included in the find. Apparently, according to the letters, Schupmann was collaborating with an optical firm, Voigtländer, to build a proof-of-principle instrument. We are lucky that these precious 107-year old documents were well preserved in Voigtländer's archives and so well cared for by subsequent private owners. It is doubtful that the instrument was ever built.

Presented here are Schupmann's drawings showing the lens cell arrangements for both the centrally-cored 175mm objective (Figure 9.24) and the 90mm

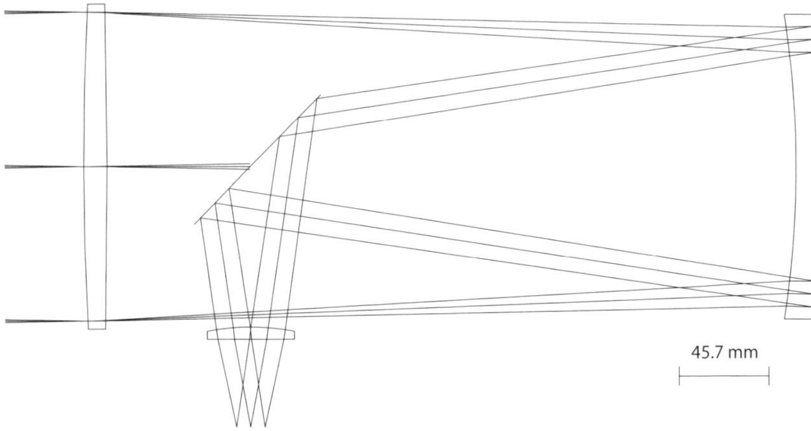

Fig. 9.23 *The Honders 6-inch f/2.7 brachymedial camera configured for CCD work by employing a diagonal mirror. As shown here the diagonal blocks 18.2% of the incoming light. The FOV is 2 degrees and fully illuminates a 15mm wide CCD chip. Of course, a film holder could also be used at the CCD location.*

back-reflecting corrector (Figure 9.25). Also included is a general arrangement drawing (Figure 9.26) of the complete instrument. The mounting is an alt-az type for testing simplicity. Lens mounting and tilt adjustment arrangements would serve well indeed for many of the brachymedial designs detailed in this chapter, most especially those of Klaas Honders. Of course, an equatorial mount is preferred, especially for astrometric work.

Note in Figure 9.24 the dotted-in special adjustment wrench for setting the diagonal prism's rotation position. The squared hole in the end of the bar fits the squared end of the diagonal support stub, thus allowing rotation alignment of the prism and maintaining it while tightening the stub nut. Plastic shims under the washers protect the lens from sharp point contact. These days, a diagonal mirror is usually preferred for its freedom from color and spherical aberration. By accurately machining the 45-degree support face there should be no need of adjustment in that plane. In Figure 9.25 we see the long extension rods to adjust the corrector tilt which can be reached while critically examining a star image at high power. Especially note Schupmann's use of two double-ended single-leaf springs to push back against the corrector's adjustment screws. The third pivot point might well be a slightly-domed thick washer. Again we see a cell design that keeps out dust. Note that the cell rim "seal" is lightly crowned to allow adjustment over the limited range needed.

An optical prescription accompanied the documents (not included here) and the old glass types specified are #48 Zink Crown for the objective and #37 Light Flint for the corrector. Although these glasses are no longer available, it is possible to find modern near equivalents and, with very little adjustment to the curves, one can reproduce the focal ratio and working distances. The design focal ratio turns out to be about 12.2, consistent with its intended low-power use. Schupmann did not include a field group to correct lateral color. As this is the dominant aberration

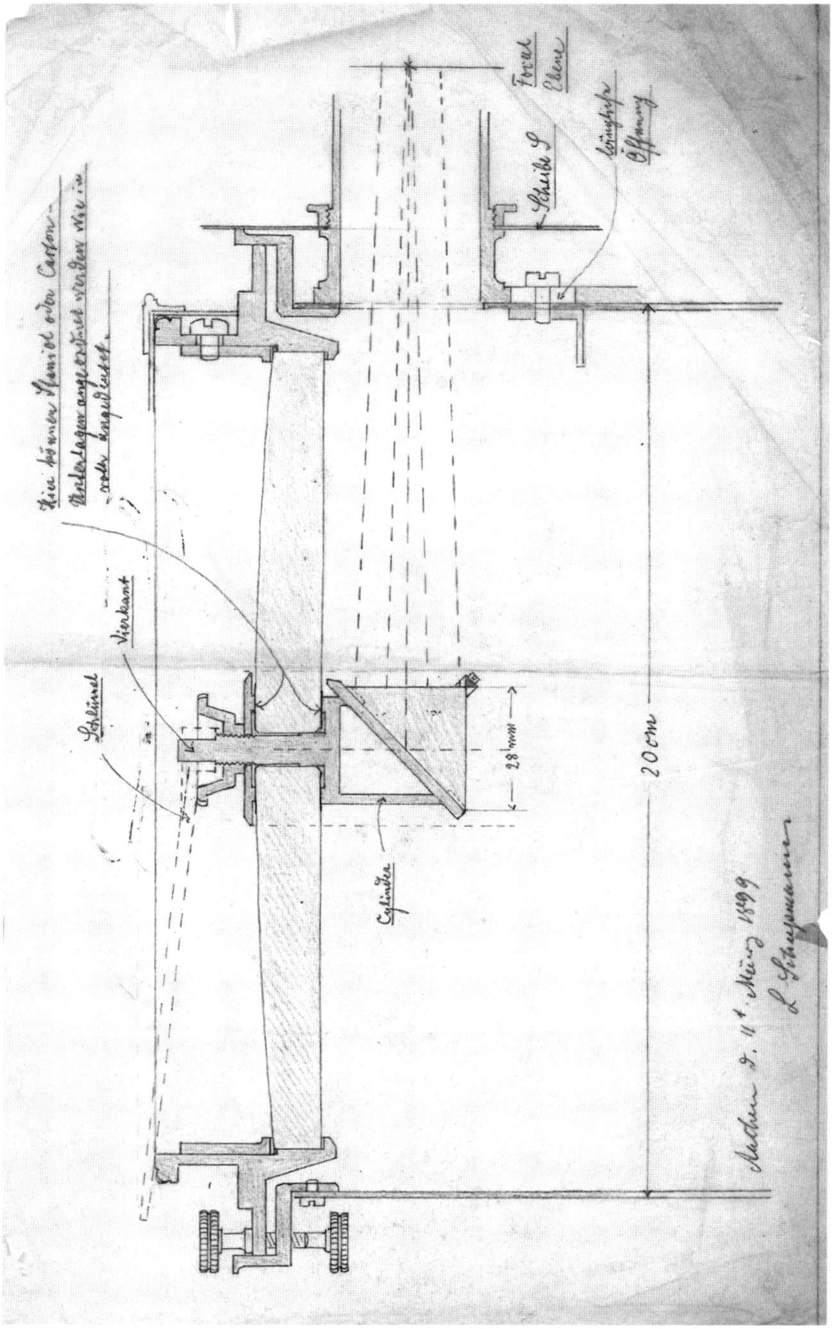

Fig. 9.24 *Schupmann's mounting arrangement for the 175-mm meniscus objective lens. Note the close fit between the cell and counter-cell to prevent dust infiltration.*

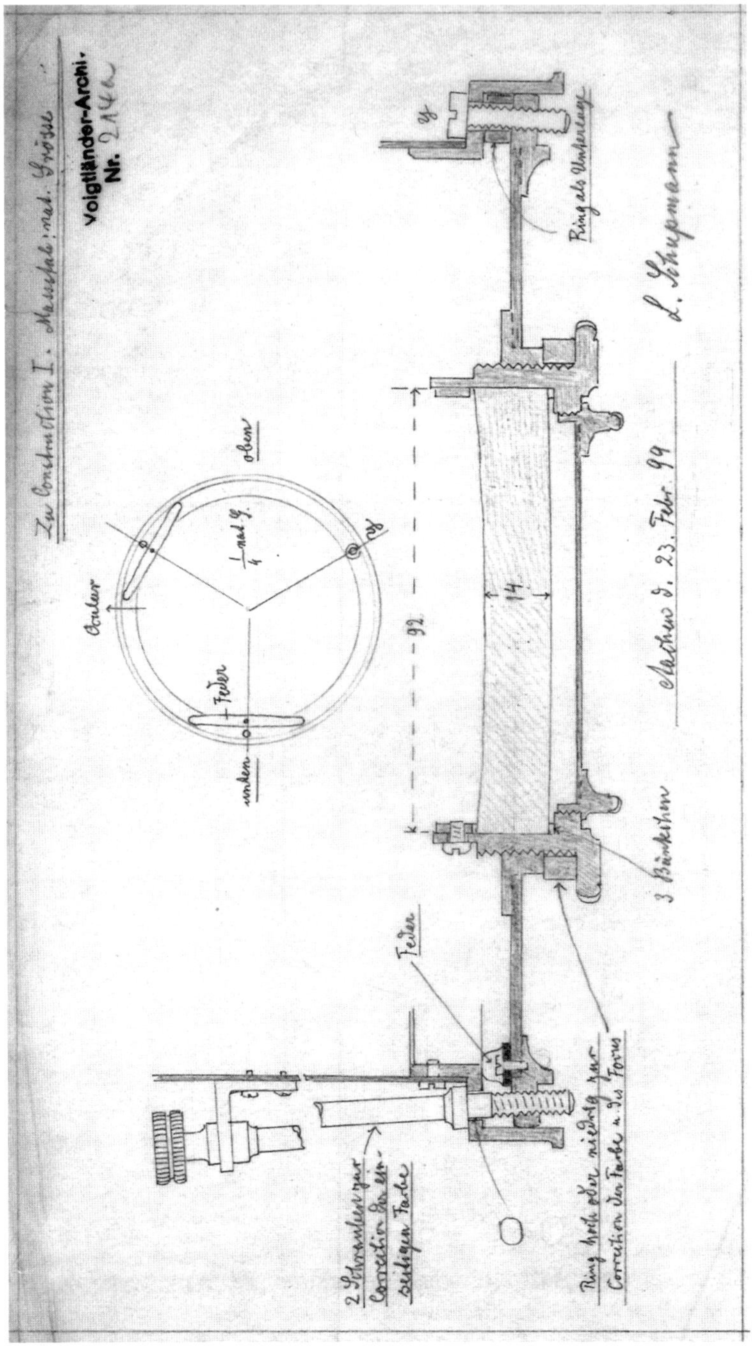

Fig. 9.25 *Schupmann's corrector lens cell and tip-tilt threaded adjustment rods. The leaf spring disposition is seen in the smaller scale sketch.*

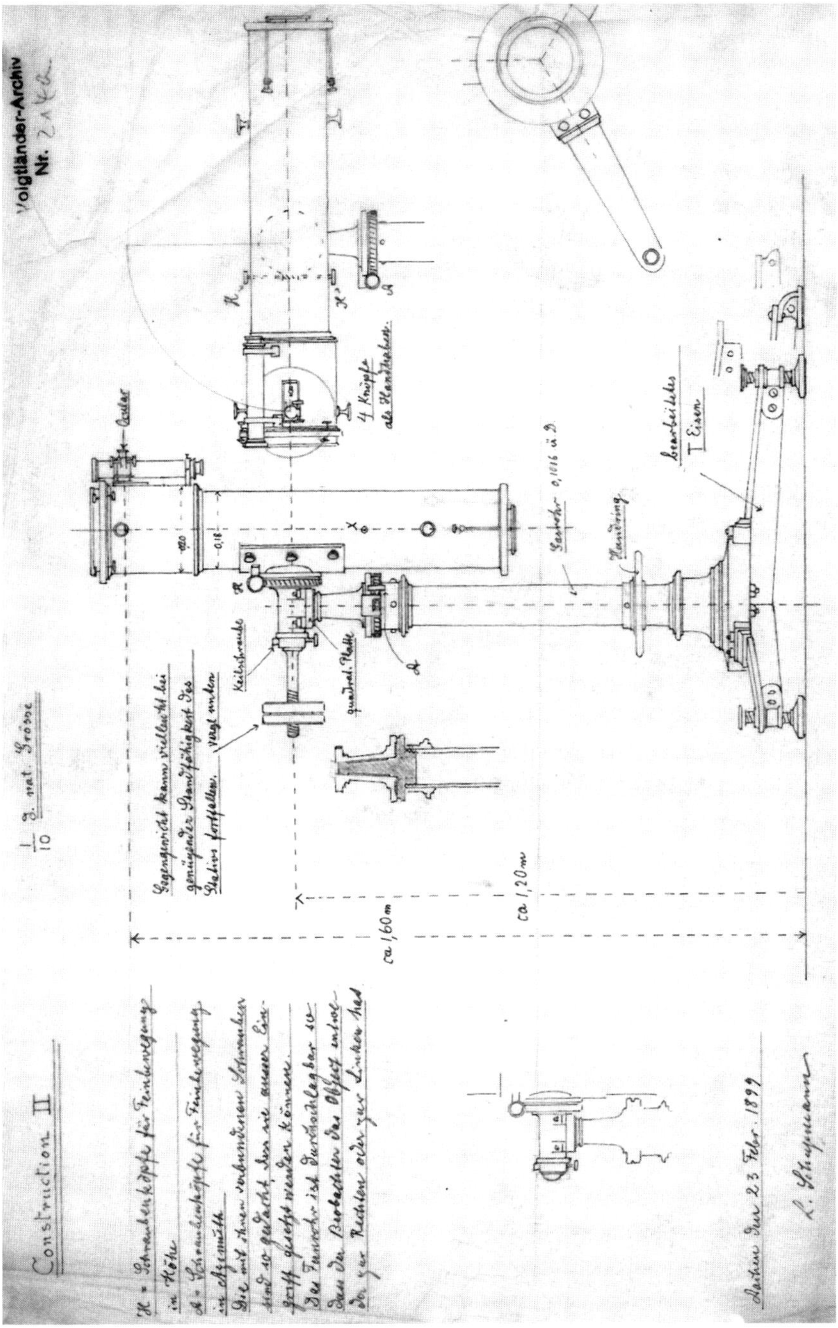

Fig. 9.26 *The proposed test mount for experimentation with the brachymedial layout. What a great comet seeker!*

in these type designs he probably intended to employ special correcting eyepieces of his own devising.

9.9 The Rathenow "Brachy-mediale"

The only traditional brachymedial known to still exist is located in the German town of Rathenow (formally in East Germany), and was built by amateur astronomer, Edwin Rolf (1899–1991) in the years 1949–1953 (see Figure 9.27). This giant, 13-ton brachy has undergone extensive restoration and is presently in use. A technical report titled "Der Verkürzter Refraktor"[4] (shortened refractor) by Rolf gives a thorough description of its design and performance, so only the highlights are touched upon here.

Fig. 9.27 *Edwin Rolf, builder of the Rathenow instrument. Photo courtesy Rainer Ortmann.*

The 700 mm (27.6-inch) BK7 objective lens is supported at the end of a 10.15 meter long steel tube. The 320 mm (12.6-inch) diameter LF-3 corrector lens is double concave and followed by a concave reflecting mirror (Schupmann's "Typus B") thus the light traverses the corrector lens twice in sending the converging *f*/19.4 cone to a diagonal mirror at mid-tube where the beam is reflected down the declination axis to a 77.5mm (3.1-inch) diameter 3-element cemented field group that corrects lateral color. This group is located about 80mm before final focus.

Although never realized, Rolf's idea was to hermetically seal the tube and lightly pressurize it using an automatic control system to exactly float the objective on a cushion of air for any tube inclination. This idea alone is remarkable as it relieves the major component of gravity-induced flexure, especially when viewing near the zenith where you want the best imaging. Rolf's study showed that this flexure, if uncompensated, has a decided influence on the achievable color correction! The finder hides in the ¼-diameter main diagonal's obstruction and actually employs the unused center of the objective to provide partial power for the 137 mm (5.4-inch) diameter finder lens which, by way of a second diagonal, sends the wide field beam to focus directly beside the high-power eyepiece. This double convergence provides an *effective* aperture of 175 mm (6.9 inches). Amazingly, these inside optical components can be accessed by removing a panel at the bottom of the tube, up which an internal ladder runs, allowing a small person to climb and make adjustments, etc.

Originally the open-air telescope and mount structure sat atop a building in Rolf's home garden. The building contained a machine shop, polishing area and

[4] Rolf, Edwin. *Abschlussbericht zum Forschungsauftrag*, F2-119, "Der Verkürzter Refraktor," Rathenow, 15 May 1953.

Fig. 9.28 *Edwin Rolf's 700 mm (27.6-inch) f/19.4 "Brachy-mediale" located in Rathenow Germany. Photo courtesy Rainer Ortmann.*

Fig. 9.29 *Rainer Ortmann shows the scale of the Rathenow Brachy by holding a meter-long stick at the objective's dew shield opening. Rainer kindly provided the author with much of the recent history of the instrument. Photo courtesy Rainer Ortmann.*

darkroom. A complete restoration of the instrument was begun in 1994 when the telescope was disassembled and transported to an indoor area for the refurbishing. Finally in 1996 the work was completed and the telescope moved to its present location adjacent to the Bruno H. Bürgel School, just 1.1 kilometers from Rolf's garden site. The brachy mounting now sits atop new, but lower piers. The objective

Fig. 9.30 *Rainer Ortmann (left) in the observing booth with Wolfgang Busch. Photo courtesy Roger Ceragioli.*

lens is still high in the sky giving good immunity to bad seeing effects near the ground (one of the real reasons refractors always show the diffraction disk so much sharper). The equatorially mounted instrument remains in the open air, although the observer is, as before, protected from the elements in a small observing booth located at the intersection of the polar and declination axes. This booth is separately supported on the steel truss-work stairway so no amount of observer movement disturbs the image stability. Although the tube presents (geometrically) a nearly symmetrical or balanced wind load, precise work requiring a stable image position is practically limited to nights of relatively light wind or calm conditions. Truly, a remarkable amateur telescope!

The instrument is used in public observing sessions and by local astronomy students. Much of the work of upkeep falls on the shoulders of Rainer Ortmann, the instrument's curator, and who played a major role in the restoration of the optics and continues to care for them.

Chapter 10
Some Important "Standard Design" Details

This discussion will emphasize the Standard Medial Design, sometimes referred to as the Super Schupmann. Naturally, much of this can be applied to any medial and a few instances are included. The standard design has three principal elements: a singlet objective, a field mirror, and a singlet corrector. The arrangements of the elements are detailed in Chapter 5. We will start with the objective.

10.1 The Objective Lens

For the standard design the objective is a double-convex singlet. Radii are chosen to reduce its coma to zero. This is easily performed in design programs such as OSLO LT 6.2, a free download version. For the recommended BK7 optical glass, and a focal length of 100.0 inches at the Sodium D line, the coma-free radii in inches are as follows:

$$R_1 = 56.648 \qquad R_2 = 587.61 \text{ , both convex.}$$

Thickness affects the focal length only slightly (0.024 inch for the case below), however, it does significantly impact the objective's back focal length (BFL). We need good knowledge of the BFL in the grinding and polishing stages, and for certain field-mirror spacing considerations. For the purpose of lens construction and testing, the BFL sets the length of the spacing test rod. This rod is cut to a specific length, which is equal to the BFL at the *test wavelength* and is used in the autocollimation test between the objective's back surface and the plane of the light source. Table 10.1 gives the standard design objective's BFL for eight wavelengths, three of which are commonly used in monochromatic testers and interferometers. The data are presented for two thicknesses (T): 0.55 inch for medials up to 6.7 inches, and 1.0 inch for instruments of 9- to 12-inches clear aperture. For other thicknesses, simply ratio appropriately. The wavelengths are in Ångström units and the BFL is in inches.

Although the values are given to three decimal places, for practical purposes one would round off to two places as that's about as close one could hope to cut a distance rod. Notice that the total color shift between 4047 and 8521 Ångstroms

Table 10.1 Table of (Objective Lens) BFL Versus Wavelength for Two Thicknesses

Lambda	4047	4416	4861 (F)	5461 Hg	5893 (D)	6328	6563 (C)	8521
T = 0.55	97.138	97.901	98.602	99.300	99.681	100.000	100.150	101.039
T = 1.00	96.894	97.657	98.358	99.051	99.437	99.756	99.901	100.791

is 3.90 inches. By locating the field mirror at the 5461 focus the out-of-focus beam patch diameters, at the spectral limits given here, are roughly the same. At *f*/10 the patches at these spectral extremes are somewhat over 0.200 inch. The field mirror can, of course, be located at any desired color with Hα (6563) one obvious choice for the solar prominence researcher using an occulter or occulting hole. The medial curves can be slightly adjusted for various choices of field-mirror positions along the optical axis; however, it is possible to change the field-mirror position and then simply shift the distance between the corrector and field mirror to restore color correction at the expense of a tiny reappearance of coma and spherical aberration. The standard design is slightly tweaked to bring the final focus about an inch beyond the plane of the field mirror, and shifting the field mirror back to the Hα focus will pull the focus in towards the tube.

A modified standard medial prescription specifically adjusted for Hα prominence observations is presented in Chapter 7. This design positions the field mirror nominally 0.080 inch beyond the Hα focus. The final focus is 1.6-inches beyond the plane of the field mirror. This gives us some wiggle room, so when the field-mirror occulting hole edge is brought to the Hα focus the resulting optical effects are harmless. Amateurs with ray trace abilities can really fine-tune their particular medial for any sort of field mirror position.

A practical aspect of the patch size on the field mirror is its impact on the maximum unvignetted field of view, which is defined by the field mirror. This is because the beam footprints at the CCD spectral limits can begin to fall off the edge of the field mirror while the median wavelength is still safely inside the mirror edge. Now for most observing this would go unnoticed; however, CCD photometry in BVRI-bands over the full field may possibly be compromised. It is vital to know beforehand what the field limits really are for photometric work. By placing the 5461 Hg focus at the field-mirror vertex we maximize the photometric FOV. For a field mirror of 1.250-inch diameter we can safely have a linear full FOV of 1.250 – 0.200 = 1.050 inches. This consideration must be strongly in mind when using a focal reducer for wide-field photometry, but will not enter the picture in photometry at the normal and Barlow-magnified focus positions. This is particularly true with amateur CCD cameras where the chip is much smaller than the unvignetted field.

10.2 The Field Element — Lens or Mirror

Schupmann's great breakthrough in perfecting the medial telescope was his introduction of a field lens. Now the corrector could lie, not before (as in the brachymedial), but after the objective's focus. By positioning his field lens at the objective's median focus, complete correction of lateral and axial color was finally possible. This is achieved by making the focal length of the field element such that the first nodal point of the objective is imaged on the corrector's second (reflective) surface vertex which, in effect, optically unites the two elements. To put it in modern jargon, beam-walk is reduced to zero, or in other words, the image of the objective on the corrector is fixed regardless of off-axis field-ray angles.

The action of a simple field lens or mirror in the medial presents to most ATMs an almost incomprehensible optical scheme, and the author finds most questions concerning the medial center on this element. Why can't the field mirror be flat, after all it simply folds the system, they ask? First, a medial employing a flat field mirror would be one of the worst, if not the worst, performing telescopes ever made. Secondly, it is not there to "fold" the instrument as any Schupmann scheme can have a tube length equal to the focal length. By definition, folding an instrument is always accomplished with a non-active (powered) element, a flat! The standard medial is *not* a folded refractor as one so often reads in the popular literature! Also, by definition, a shortened refractor employs an optically-active element as for instance in the brachymedial described in Chapter 9. Your photographic telephoto lens is also a shortened refractor.

10.3 The Polish

Lenses or mirrors that lie at or near an intermediate focus require the best possible polish. In the Schupmann design the slightest residual "grey" in the finished surface is very bad indeed. The patch of light for each color has a different diameter at the field element yet at the final focus all colors are united in pinpoint star images. However, the corrector still softly images the original patch if it is grey (a sort of new source), giving a haze about the perfectly corrected final image. Technically we are talking about the non-specularly reflected (scattered) component. This scattering component is, unhappily, strongest in a narrow cone about the specularly-reflected rays of a field mirror, or transmitted axis in the case of a field lens. To bring this scattering to an acceptable level you must completely polish out the field element whether it is a lens or mirror. Surface roughness on the order of two nanometers is easily attained with careful polishing. For the field mirror, the forward, in cone, scatter of 1/10,000 of the specular beam is achievable at this level of polish even after coating with protected aluminum. Thus, the integrated haze, covering and surrounding an object, is down by ten stellar magnitudes. Because it is spread out its surface brightness is much less. A field lens will exhibit less scatter even when anti-reflection (AR) coated. The best test of polish is to focus the Sun on the field element surface(s) with a short-focus lens. A common eye loupe is perfect for the job. This test will reveal the grey by showing the image of the Sun on the surface in question. Use a black background and fiddle with angles such that your eye is near the specular rays and get a big surprise! Incidentally, this test for polish should be passed by all the Schupmann elements for best overall results. The field element is so small that a perfect polish should only take about two to three hours if your fine grinding is complete.

10.4 The Radius and Figure

Getting the radius right-on with a good sphere is a bit of a task. However, by flipping to "tool on top" the radius is easily lengthened or vice-versa in the polishing stage. If the first test-polish to check radius is off by an inch or so it's best to correct this in grinding to within ¼-inch. Except in the case of a medial employing a full-size corrector one may think the figure on a field mirror should be an ellipse to, perhaps, more sharply image the entrance pupil at the working conjugates. To the author's knowledge, this sort of refinement has never been tried and at the *f*/ratio of the tiny element, would be hard to detect in the Foucault test. One could, none the less, null figure the element at the working conjugates after getting the radius adjusted.

10.5 A Field-Mirror Turret

For combined solar and night work a field-mirror turret should be seriously considered. In the mid '60s Edwin Olson proposed making the field mirror way oversize so different portions of the surface could be rotated in position thusly: position 1 would be aluminized for star work, position 2 would be left un-aluminized for white-light solar work and positions 3 and 4 would be bored to just clear the Sun's focused disk diameter in Hα for winter and summer, respectively. These bored "occulting" holes would pass the heat and light of the Sun harmlessly out the back of the scope, with their remaining surface (aluminized) reflecting the prominences. This occulter idea has been tried with perfect results by the author and other Schupmann enthusiasts. The turret needs two focal positions, one for night work at the design location and another farther from the objective at the Hα focus. (Chapter 7 describes a design with the field mirror at the Hα focus making a two-position field mirror unnecessary). The upset color correction is of no concern when viewing prominences, as the filter is far too narrow to reveal it. The disk for the field-mirror turret must be optical grade quartz polished flat on the back and must be wedged about 4 degrees, so that when used in a white-light solar mode the second surface reflection misses the corrector. (Don't worry about the reflected energy (4%) from the second surface heating the flat black walls of the tube and baffles—it's too weak.) Of course, one could design the wedge angle to reflect the back reflection out the objective lens! The entire turret needs the normal tip-tilt adjustments of any field-mirror system.

10.6 The Field-Mirror Cover

The field-mirror cover serves two functions: it acts as a dust cover and it provides a support for an alignment cross hair. It is vital that the field mirror be covered when the medial is not being used, as dust will otherwise accumulate on its surface. Other than solar prominence work, this is of little concern on the objective or corrector as the contamination is in pupil space and is thus not imaged. The field mirror, on the other hand, lies at an intermediate focus and the corrector re-images the dust specks at the final focus. This looks, at first glance, to be a big drawback to the medial design. On closer examination a far less frightening picture emerges. This is because the field mirror is only in focus over a very narrow spectral range as it is located in the region of uncorrected color. This primary axial color spread is 3.0 inches between 404 and 656nm (visual) and 5.3 inches between 360 and 900nm (CCD) for a 100-inch FL system. Large dust specks are therefore small compared to the average color blur diameter, and thus show themselves as weak diffuse images at the final focus, tinted in the color at which the field mirror is positioned. An easily removed side plate on the tailpiece will provide access to gently brush away any accumulated specks. After using medials for forty years the author has rarely found it necessary to dust the field mirror.

The cover can take many forms; however, it must be simple to open and close without removing a side plate. This is best accomplished with a rotary shaft extending through the support "box" structure. The idea here is to seal the tailpiece from dust infiltration as much as possible. Although the focuser still presents an opening during brief changeovers, we can reduce the problem to a minimum by always plugging the drawtube after each observing session. One of the best cover designs is a simple flip-type that swings over the mirror. An example of this is shown in the photos in Figure 10.1. In the top photo the shaft carrying the cover can be seen extending through the tailpiece box as a knurled knob. Note the swing cover on top that closes the diagonal mirror port as well. The focusing stage, which holds the CCD camera, auxiliary lenses and BVRI filters, is totally divorced from the field-mirror box. See Figure 10.2 for the total picture. Another arrangement has the cover integral with the field-mirror cell. This design (see Chapter 11) keeps the cover square regardless of mirror adjustment as it goes along for the ride. As a practical matter the flip-type cover can be squared-up after the lateral color adjustment phase is over, and will remain dust-tight for years.

A cross or other mark must be scribed on the cover and must lie directly over the center of the field mirror when the cover is fully closed. This cross can be seen, rather blearily, in the eyepiece at the final focus and is centered in the field by tweaking the corrector tip-tilt adjustment. The scope is pointed at the daytime sky with the objective cover off for this work.

10.7 The Field Lens and Cover

In-line and various elbow arrangements such as Kutter's employ a field lens to image the objective on the corrector. Occulting cones and masks are easy to implement with in-line set-ups. Many people ask about correcting the field astigmatism with a specially shaped field lens, but this is an optical impossibility. (See Willard's anastigmatic medial design in Section 6.3 on page 68 to appreciate this). Some improvement in lateral color correction is seen in tiny corrector systems by achromatizing the field lens; however, in practice the author has found that a single-element lens works fine. Even bending the lens shows little difference in ray trace analysis, with only the slightest (geometric only) improvement when the ra-

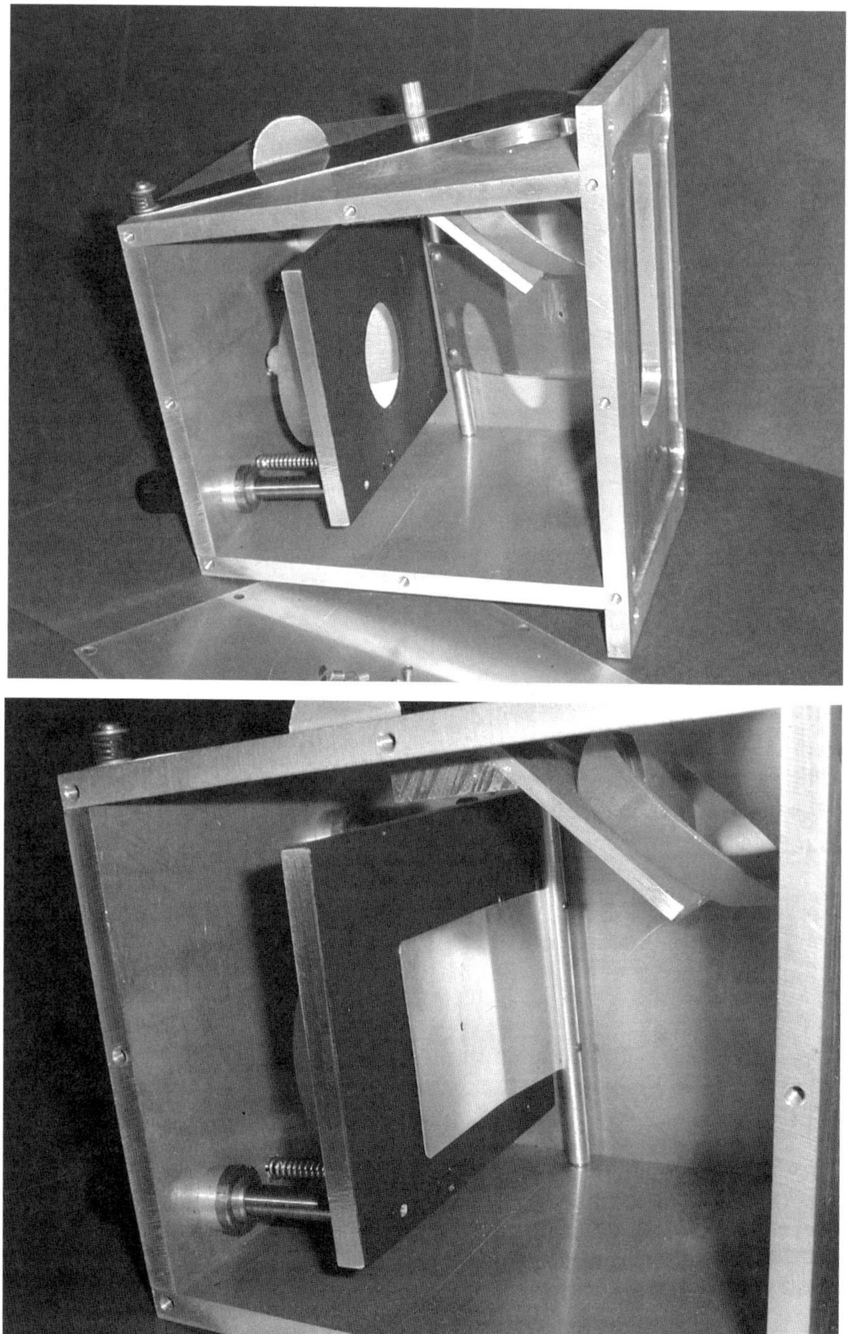

Fig. 10.1 *The author's 9-inch medial with the field-mirror cover open (top) and closed (bottom).*

Fig. 10.2 *Tailpiece of author's 9-inch medial.*

dii are proportional to the working distances and the *shorter* radius faces the objective. A plano-convex lens is about as good. Naturally the polish and focal length requirement are the same as for the field mirror and, of course, should have a similar cover arrangement. Remember that now the alignment cross faces the field lens! By making the cover of clear plastic, the cross is easily seen with the objective open to the daytime sky and will not require any sort of flashlight illumination during the alignment phase. By parking the scope pointing up it is only necessary to cover the field lens' skyward face. A long (2-inch) hood could provide dust protection on the lens' downward side. A very practical addition to the tailpiece is a heater to keep the temperature of the structure above the dew point when the scope is not being used. (The heater will go a long way toward keeping the entire telescope dry). This will prevent pinholes from forming in the field-mirror coating almost indefinitely. Small, 20-watt, bulbs located inside and out of the ray path do the trick. Just remember to turn them back on after an observing session. It is by paying special attention to these and all other details that your medial will perform at its theoretical best. No shortcuts allowed!

10.8 The Corrector

This little meniscus critter provides lots of construction fun. The curves are adjusted to give the same axial color as the objective when the corrector is used at one-to-one conjugates, but because the refraction is negative the color sign is reversed,

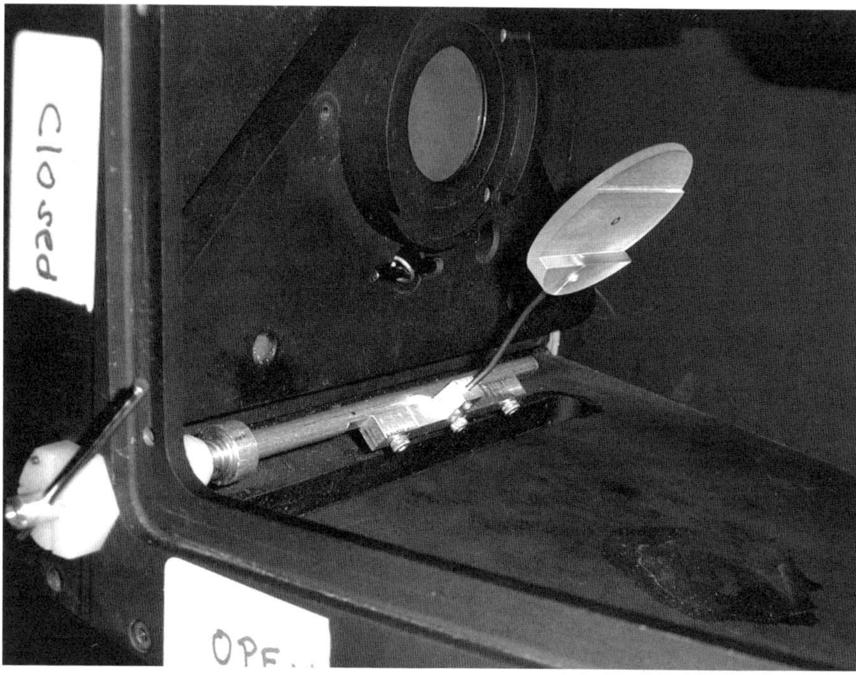

Fig. 10.3 *A field-mirror cover at mid-travel. 13-inch Stellafane Medial. Note alignment dot.*

exactly compensating the objective's positive axial color. The back surface is reflective and serves only to return the rays upon themselves. If we test its focal distance in Sodium light our radius (distance) rod would be 54.59 inches. For an Hg tester we would cut the rod longer by 0.386 or 54.976 inches. Conversely, if testing in Helium-Neon light (interferometer) we would cut the rod shorter by 0.319 or 54.271. It is assumed that your knife and source are par-focal. If they're not, compensate accordingly!

One of the most challenging operations is the polishing of the corrector's back surface. Because the reflection takes place in a high-index medium the wavelength is shortened by the index value, thus 1/20 wave on the surface takes on a new meaning. To attain an effective 1/20 wave surface, we need to keep the surface sphericity to within 1/30 wave or better. This will give a final wave-front error of 1/10 wave. Because we correct the system in autocollimation a weak, smoothly varying back surface error can be touched-up on the refraction surface. It is the smoothness that is very critical on the back surface as micro-ripple negatively impacts the Strehl ratio ~32 times more than the same ripple on a single refracting surface. If the Schupmann has a "Achilles heel" this is it, so, builder pay attention to this surface and be sure it is as smooth as you can make it.

Chapter 11 Schupmann Construction Details

11.1 Introduction

The author realizes that building a medial is hardly a beginner's project and yet, those amateurs with advanced skills need little more than an optical prescription to complete a fine instrument. Naturally, the question arises whether on not to include construction details in this volume and if so, to what level of detail. Being primarily a refracting telescope of a highly unusual nature, it was felt that important points of construction should be included. Particular emphasis is placed on the final stages, which include alignment adjustments, associated devices, and tweaking for best star performance. It will be instructive to also read Chapter 13, by Scott Milligan of the Springfield Telescope Makers. He sparked interest and provided the leadership in building the club's 13-inch medial located in the McGregor Observatory. (See Figure 12.1 on page 155.) Although this book emphasizes the "Standard Medial Design" much of what follows can be applied to any of the configurations shown. The standard design, being all-spherical, will require only minor figure touch-up in the autocollimation phase. This is not a design failing, but due simply to the extreme difficulty in producing essentially perfect spheres. This is particularly true of the back surfaces of both the objective and the corrector. One could, in principle, make full-size spherical test plates for all convex surfaces and obtain an excellent wave-front correction. Stellafane's Schupmann was done that way. The fabrication technique described here will produce the same correction by using only one sub-aperture spherical test plate matched to the first surface of the objective. Most of the instructions in this section apply equally well to non-spherical designs. These are useful for those building a medial with a small corrector. Also, if solar prominence observing is envisioned and you plan on using a standard (old fashioned?) occulting cone, then the instructions on null figuring the objective must be followed unless your design focal ratio is *f*/16.6 or higher. This is because a spherically corrected image is required at the occulter for substantially faster systems (see Chapter 7). So, let's get started.

11.2 Glass Purchase and Specification

This information is to be applied regardless of the size instrument you decide to build. Most designs detailed here employ Schott's BK7 optical glass. This glass has very good physical properties for a fine medial. Among them we count the following: relatively low cost, ability to take a coronagraphic polish, fast polishing, low internal scatter, and thermal characteristics which tend to maintain the system EFL. Other glasses are similar such as Bausch & Lomb's BSC2, and can also be used. Even an optical grade quartz system is possible, as is shown in Section 5.6 on page 61. In the early days of medial development in America we had a "leave no stone unturned" philosophy and thus insisted the blanks be from the same melt. Smart ray trace gurus have pointed out to the author that the differences in partial indices of random BK7samples will still produce a Schupmann optically indistinguishable from a same-melt instrument. This has proven to be true in practice as the author's latest medial demonstrates. However, one must be careful not to use glass of unknown type and quality found at swap tables, etc. Unless you're really sure about the origins of a particular set of blanks; buyer beware. One rule is to insist on the manufacturer's *test certificate*, after all, how do you know that, for instance, a partially worked blank is really BK7? The test certificate will state the blank diameter in mm as a clue. More important, the certificate gives the builder vital data concerning the optical quality, including the refractive index, partial indices, birefringence and homogeneity, striae, etc. These data should show homogeneity values on the order of 5×10^{-6} or better. When you order the blanks explain their use and ask for minimum bubbles (hardly ever a problem these days). The cost of same-melt glass is the same and I still advise the builder—if ordering (not scrounging) from a respected supplier—to request "same melt." Treasure your test certificate that comes with the glass! The Schupmann medial is a difficult but rewarding project, so don't blow it from the start!

The glass should be ordered edged to a specific diameter and thickness. The advanced amateur will know his typical stock removal to achieve the desired finished thickness and will order accordingly. For the standard all-spherical design order the objective about ¼-inch over the desired clear aperture with a thickness ratio of 10 to 1. Order the corrector blank at least 10% larger than its clear aperture would indicate. This will allow for the inevitable turned edge and the diverging rays within the element. A blank thickness ratio of 5 to 1 is usually fine for the corrector. Naturally the builder should draw his elements to full-scale to get an idea of how they will come out of the blanks, thus indicating stock removal, before ordering. The corrector is easily adjusted for thickness errors and the objective BFL is only slightly changed. If you think you can make a really thin objective successfully, than go to it as the thermal settling time is greatly shortened and the scope weight comes down. It will, however, dew-up sooner!

11.3 Test Equipment and Materials

11.3.1 Optical Flat

Complex optical systems, especially those of a refractive nature, are almost always tested in autocollimation and the Schupmann is no exception. This test requires a highly corrected optical flat at least ½-inch larger than the objective. Often it is possible to borrow a fine flat for the project. Make sure of the optical quality, 1/20 wave being a minimum accuracy.

11.3.2 Knife-Edge Tester

The Schupmann can be figured in white light by testing the whole system in autocollimation. If you plan to use the instrument for solar prominence observing then null figure the objective in autocollimation with a monochromatic source. The corrector should be figured in a total system test in white light. It is almost essential that the source and knife-edge are optically the same distance to the element under test. This makes the use of radius (distance) rods much easier. Laser interferometers are part of many advanced amateur's test kits and are the ideal devices for most of the work, especially the final system correction tests. A vibration-free test bench will need to be constructed for interferometry. Since a test bench is needed in any event, the extra step to obtain vibration isolation is really worth it. (See Section 13.7 on page 184).

11.3.3 Filters

Obtain a green filter or filter combination which will give a roughly 40nm band pass centered on 546.1nm You will need this filter to test and adjust the focal length of the objective and corrector. If you have a mercury light source for your knife-edge tester this filter will isolate the 546.1 nm line. A good filter combination is Kodak's #58B and 77A. If all your testing is with a helium-neon laser interferometer then the focal test distances will need to be recalculated for 632.8nm when measuring individual elements. A Kodak #92 (deep red) is often useful in setting up a solar prominence scope for Hα work. Many builders are currently using laser diode modules (LDMs) in their knife-edge testers. They are bright, monochromatic and available in a wide range of wavelengths.

11.3.4 Test Plates

Only one test plate is needed to complete the optics. It is concave and is used against the objective's first surface. The plate should be of 4-inch diameter for instruments up to 8 inches CA, and 6-inch diameter for 10- and 12-inch projects. The test plate is ground and polished to a ⅛-wave sphere and the back polished enough to view the interference fringes when placed against R_1. The radius of curvature of the test plate should be held as close as possible to the design prescription.

11.3.5 Sodium Lamp

Some sort of line source is required to obtain fringes with the test plate. The author uses a sodium lamp, but a small inexpensive neon bulb shining on a diffuser works just fine. Most advanced amateurs have nice test boxes to support the parts such that a 45-degree windowpane allows a good long viewing distance.

11.3.6 Radius Rods

Five radius or distance rods must be fabricated. One is used to measure the test plate radius; two are needed for the first surface and the conjugate focus of the corrector. The fourth is for adjusting the back focal length of the objective. The last is for the field mirror. These rods are made of aluminum or hardwood. They are cut to an exact length equal to the BFL or the radii indicated in the design specification and in the case of BFL, adjusted for the test wavelength. As mentioned earlier, the knife-edge tester must have the source and knife-edge in the same optical plane. When you can just fit the rod between the knife-edge and surface under test then your radii (or focal distance) are right on. Naturally there is a tolerance value and this is especially true with the Schupmann, as simple spacing adjustments can make up for curvature errors—up to a point. A computer study by Scott Milligan showed a fairly relaxed tolerance, on the order of 1% and still recovers fine, color-free images. The author gets his curves to one part in 2000 with little effort. In use, the radius rods should be supported in enough places along their length to prevent excessive sag. The author suspends the rod by threads from the ceiling rafters. This method allows one to adjust them gently against the surface under test with a slight positive force by pendulum action. If you use wood, (the grain wise expansion ratio is extremely low for wooden rods) it doesn't hurt to moisture proof them. About ⅜-inch dowels are a good size for this purpose. To be definitive, the ends are tapered to a slightly rounded point.

11.3.7 Measuring Tapes and Scales

Obtain the best measuring tapes and scales you can afford. A tape of 120-inch length is sufficient for most instruments. Good tapes, such as a Lufkin 16-foot, will allow you to work to 1 part in 2,000 on the long radius measurements. For short radius measures, obtain a 24-inch Lufkin or equivalent scale. With an eye-loupe you will be able to read to ½ a line width, typically 0.0025-inches.

11.3.7.1 Spherometer

To get the curves approximately correct, a spherometer is required. Enough has been written about these devices that it should be no trouble making one. A ring type of about 4- to 6-inch diameter is large enough.

11.3.7.2 Calipers

The lens thicknesses are not critical. The objective can be read well enough with wood-turning calipers. Most advanced telescope makers will have a dial indicator device to measure thickness. This simplifies the thickness measure of the corrector.

11.3.7.3 Micrometers

A zero-to 1-inch micrometer will suffice for instruments up to 8-inch aperture.

11.3.7.4 Grinding Tools

Two full-size tools are required for each surface. One tool is for grinding and the other for the pitch lap. Try to get your grinding tools faceted after #240 grit is finished. It is necessary to curve-generate the pitch lap tools except R_2 of the objective which is, for most systems, rather flat. The builder may wish to have his lens blanks and tools diamond-generated to curve by a professional supplier. One would then start grinding with #302 emery, or its equivalent, removing every trace of generator swirl marks (fractures) before going on to the remaining fine grinding stages.

11.3.8 Miscellany

Naturally you will need grinding grit, polishing pitch, polishing rouge, and detergent to help in the grinding. These sorts of things are well known to anyone tackling a medial. The author prefers Barnesite for the polishing and elutriated Barnesite for final figuring. There are some rumored objections to using this polishing agent on health grounds so you decide. The author is still alive!

11.4 Surface Radius Tolerance

Although the medial is very forgiving of radius errors, there is a certain pleasure in working accurately. OK, perhaps it's a state of mind, but discipline started here will carry right through the project. It is the care and attention to detail that separates the genuine TN from the wannabe! The objective and corrector radii should be accurate to plus or minus 1 part in 2,000. This tolerance is not as difficult to achieve as may first appear and a first-class optician would have little difficulty working to 1 part in 5,000. That level of accuracy is clearly unnecessary.

Let's take, for example, the test plate for R_1, using the standard design nominal value of 56.648-inches. A $1/2{,}000$ tolerance is 0.0005 x 56.648 = 0.028 or just under $1/32$-inch. The radius rod can be made to an accuracy of ± 0.010 and the plate can be polished to within ± 0.010 of the rod (assuming a constant temperature work room). Matching the objective to $1/2$-ring keeps the desired curves well within tolerance, assuming the errors do not all add! The shorter radii have correspondingly smaller dimensional tolerance; however, the optical sensitivity of the test is much higher. Also the short radii rods are measured with the engraved scale. Some builders may have access to a tall vernier height gauge to get the short R4 (corrector concave side) rod dead-on. *These tolerances can be much looser if one is willing to compensate the errors by various spacing adjustments in the final autocollimation tests; however be sure the errors tend to push the final focus away from the tube or image access will be compromised.* Scott Milligan has made an error analysis of the standard design and finds no image degradation (after spacing compensation) with radius of curvature errors of ± 0.1% or 1 part in 1000.

11.5 Thickness Tolerance

The central thickness of the objective and corrector can be any reasonable value you choose consistent with minimum tolerable element stiffness at one extreme and thermal settling time at the other. The standard design thicknesses are simply nominal values. Advanced amateurs will have entered their medial's prescription, with their chosen thickness values, in their favorite ray trace program making it an easy matter to adjust both the corrector's back (convex) radius as the final thickness is approached and the test rod length for the objective adjusted based on expected final thickness. If building right to the prescription, thickness errors on the order of ± 0.1 inch can be ignored for the objective, but the corrector should held to ± 0.030.

11.6 Getting Started

We will divide the optical work description into three sections: the objective, corrector and field element (lens or mirror). We will take up the spacing and alignment of the optics later.

While waiting for your glass to arrive, work can begin on the test plate for R_1. You will surely have some plate glass disks in the scrap box for this little project. It is not necessary to obtain a perfect polish, just get the radius correct with a good ⅛ wave spherical figure. If you used plate glass the back is clear, otherwise the back should be done first. Be sure to allow plenty of settling time between spells of radius adjustment. When the radius is within ⅛-inch, after repeated fine grinding and quick polishing spells, you can bring the plate to tolerance by polishing tool-up or tool-down. And this is the way it will go for all the surfaces, so get used to the idea of adjusting the radius with 303½ emery and then a quick polish followed by an optical test. Use your spherometer to up to 303½ emery then put it back on the shelf where it will cause no harm!

11.7 The Objective Lens

Various Schupmann configurations require the objective to be aspherized. As mentioned before, if you plan to use an occulter for solar prominence observations the objective must be aspherized. Another configuration where this is true is where the corrector is same diameter as the objective, or "full-size systems". This is because full-size correctors, by themselves, are free of spherical. For the all-spherical standard design one would polish the objective carefully using strokes that promote a spherical surface. You will get some help by using the test plate to check the figure center compared to the edge. With care good starting point spherical surfaces will result. If the design calls for a spherically corrected objective then use your green-filtered Hg tester and bring the lens to a perfect null in autocollimation. The author's Hg tester is shown in Figure 11.1. A petal lap used on R_2 is the author's favorite method for this work, thus if re-polishing was ever required due to weathering or damage of any kind to the first surface, the aspheric

Fig. 11.1 *The author's monochromatic tester uses a mercury pen-light, which is imaged on the slit with a microscope objective after which a tiny 45-degree mirror directs the light to the lens under test. The aluminum housing holds the special filters to isolate the 546 nm green line.*

would have, most likely, remained protected and undamaged.

Quickly running through the grinding steps: when your blanks arrive, the first step is to get a good 45-degree chamfer (bevel) on the edges; next, taking the objective blank, remove wedge, if any, by grinding on the side to become R_2. Be sure to first grind the tools together on *both* sides to flatten them so when they're used on the lens blank they'll make good contact and not rock when tool-down. Keep the tools heavily chamfered and keep a 1⁄16-inch chamfer on the lenses at all times. Never let it go below that value! The shallow back surface of the objective can usually be brought to curve with #240 emery in an hour or two depending on the design. However, first bring R_1 to curve (the author usually measures the concave tools) with #80 carbo, controlling wedge till #240 emery is reached. Now bring R_2 to curve with your spherometer. Bring both sides along together keeping wedge to within 0.002. When working lens-down, always use a tool with a complementary curve as a "nest" for the opposite the side of the glass that you're currently working. Tape this tool and lens around their edges to hold them together. Use electrical or wide masking tape. Use very thin paper to safely space the elements.

Wedge tolerance is not tight on a Schupmann as the system is optically centered in the alignment process. When 303½ is reached put a quick polish on R_1 (remember we must have separate polishing tools) and check with the test plate. If the curve is within 20 rings after finishing 303½ move on to the final grinding powder adjusting R_1 to obtain a one-fringe match. Polish R_2 enough to autocollimate the lens using your monochromatic tester. Make a radius rod to match the design BFL for the objective in green light and adjust R_2 to match the rod to within ⅛-inch. Use the center zone for these tests as considerable spherical will be present. If your finished thickness comes out different from the design, the BFL can be recalculated using the following formula: $\text{BFL} = F - t\,(\text{N} - 1)/\,\text{N}\,R_1$ where F is the objective focal length, t is the finished central thickness, R_1 is the first surface radius and N is the refractive index. Usually the finished thickness ends up about 0.020 less than that measured at the completion of 240 carbo or emery. This rule also works well for diamond-generated surfaces. Finish fine grinding on both sides and completely polish R_1 to a ½-fringe match (half-size test plate). Do your best to get a spherical surface by feel. Next polish out R_2 adjusting the BFL all the while. For all-spherical systems simply polish both sides, with lots of cold pressing between spells keeping the BFL right on. Reasonably spherical surfaces will result. For systems requiring the objective be spherically corrected this is the time for that work. Get a good null cut-off and the objective is done. Remember that final figuring is done in white light in autocollimation for all systems, and for the all-spherical design, a figure touch-up can be done on the corrector's first surface even after its back is silvered.

11.8 The Corrector

Grind the concave side of the corrector to spherometer accuracy stopping at a #120 carbo stage. Grind the convex side to the design radius and ~0.040 over the design thickness (fine grinding allowance). Control wedge to ±0.001. Continue through the grits until #302, making spherometer corrections to R_2 by measuring the concave tool. Grind both sides with #303½ and polish R_1 enough for a knife-edge test. Adjust R_1 using your accurately cut radius rod as a gauge. Incidentally, it is a good idea to coat the ends of metal radius rods with nail polish to prevent accidentally scratching the surfaces. Now polish R_2 and set up your Hg tester, (or white light with filter), and null the focus of the corrector. The focal length of the corrector is conjugate, and is reckoned from the light source to the vertex of R_1. In the standard design this distance is 54.59 inches. Using a radius rod cut to the focal length of the corrector *at the test wavelength*, see Section 10.8, adjust R_2 until a good match is obtained. Finish # 305 and polish out R_1 keeping it right on radius with a good spherical figure. Polish out R_2 using your radius rod to keep the focal length correct, which, as you see, also brings R_2 automatically to curve. Do not disturb R_1, do all focal length corrections on R_2. If you have a monochromatic tester and the design requires this element be corrected for spherical, you may now aspherize R_1 to give a no-zone cutoff. If you plan to aspherize the system in white light, you must first grind and polish the field element. In any event, the field element is the next job on the list so let us proceed with the optical component that is so simple by itself—but without which the Schupmann would fail.

11.9 Field Element

The field element (lens or mirror) optically superimposes the objective onto the corrector. If a field lens is used, provision must be made in the test setup to smoothly translate the lens in x and y. If a field mirror is used, provision must be made to tip and tilt the mirror. A lengthy discussion of the practical details of the field element is included in Chapter 10.

11.10 Field Lens

A field lens system is recommended for combined solar prominence and planetary work. A single element plano-convex lens of top quality BK7 or BSC2 is fine. The position of the lens in the system is best when just beyond the Hα focus of the objective. This will provide enough space to slide an occulting cone just in front of the field lens. Thickness has little effect on the focal length of a weak plano-convex lens but do not make the lens any thicker than necessary. A diameter thickness ratio of 8:1 is good. The clear diameter of the lens will determine the field of view possible. We advise a lens of 1.25" clear aperture for all systems. This diameter field lens, in the standard system, gives about a 0.75-degree field. After the lens is fine-ground to rough curve give it a fast polish. Set up a source of light with a light red filter. Set the field lens at a distance away form the source equal to the focal length of the objective, with its convex side facing the source. Place a white card at a distance equal to the corrector focal distance on the image side in a line with the lens and source. Adjust the lens radius until a sharp image of the lamp filament is obtained. Work the lens to a good sphere. There is no need to figure the lens for zero spherical at the above conjugates. More important than exact focal length or figure is the polish; it must be black. Note that the Hα focus for the Schupmann objective is roughly 1.0075 times the (Hg) focal length.

11.11 Field Mirror

The field mirror is straightforward and should only take a day or two to complete. Use a very low or ultra low expansion material such as quartz, Zerodur, ULE, Cer-Vit and Astro-Sitall. Small blanks of this material can often be found at convention swap tables and edged to size. Make it 1.25" diameter (1.5-inch blank) and give it the best polish possible. One plan might be to make a 3-inch diameter field mirror and cookie-cut three 1.25-inch mirrors from the finished piece. This makes radius control much easier and gives spare back-up mirrors, thus no down time for re-aluminizing. Use a radius rod cut to match the design prescription. Figure the field mirror to a good ⅛-wave sphere. Again see the special discussion on the field mirror in Chapter 10. The field mirror radius specification is insensitive to changes in the objective BFL (due to a thickness variation only), however, its spacing is.

11.12 Bench Alignment and Final Figuring

In order to do final testing and figuring the field mirror and corrector back surface will need to be coated. You could use three approaches: coat with a silver wash for this phase and do final coating when finished; send out for final coating; or wash coat the corrector and final coat the field mirror. The field mirror or lens is the one element that is the easiest to be certain of. It is possible to test the system with just the field mirror coated, but the amount of light returned is a little low for revealing subtle surface defects. Construct temporary wooden cells, which will support the elements at a common center height above your nice new test bench. It is this exciting phase of the project that always holds the greatest interest for the author. For the first time you will test the figure, but the eagerness to see a perfectly color-free image of the returned incandescent source can't be described! Back to the cells: provide adjustments as follows:

- **Objective** The objective will need an adjustment screw to tilt it, in a plane, 90 degrees to the corrector tilt. Because the elements are held at a common height, this adjustment is an elevation tilt.
- **Field lens** A field lens needs to be finely adjustable, vertically and horizontally, perpendicular to the optical axis.
- **Field mirror** A field mirror will need tip and tilt adjustment so arranged to work in elevation and azimuth. It's probably best to take the time and build the final cell as detailed elsewhere. This whole tailpiece assembly is one of the most technically demanding mechanical parts of the medial. See Figure 11.14 that shows the author's field mirror set-up to get some ideas. Also read carefully the special field element section.
- **Corrector** The corrector will also need tip and tilt adjustments. It's desirable to tack rails to the bench to guide the cells when roughing in the spacing. The cells have side "feet", sort of like short skies as in the example for the objective test cell shown in Figure 11.2 Arrange to fine focus the corrector along the axis for tuning out color. A ¼-20 bolt threaded into a heavy block on the bench can push the cell along smoothly enough.
- **Optical Flat** Position the *aluminized* flat near the objective. Its cell needs fine tip and tilt control adjustment for autocollimating the system.

If a field mirror system is being tested and you allowed space for the focused beam to be examined right beside the field mirror, then no additional optical element is needed for final figuring; however if you are testing a system employing a field lens, then a small flat (diagonal) is required to bring the focus out to the tester.

Set up a bright white-light source such as a 6-volt clear glass bulb. Locate the bulb beside the field mirror 1.25-inches on centers and at the centerline height. Make a little blackened tin shield so the light shines mostly toward the corrector,

Fig. 11.2 *Autocollimation test setup for the author's 9-inch medial's objective. The simple wooden cell allows the lens to be easily lifted out for the next figuring spell. Note the "runners" and wood stick guides.*

thus shielding your eye. Use your corrector radius rod to initially set the bulb to corrector distance. Now aim the corrector so the bulb's filament image is centered on the field mirror (or field lens). A paper centering target taped to the field-mirror such that it can be flipped away is very important as for all the next tests it is vital that the optical axis intersect the field mirror center, and the 1.25" center-to-center distance be maintained. There will be more detail on this later. Tilt the field mirror to center the beam on the objective. The distance from the field element to the objective should be already set with the BFL radius rod. One point to take care of is setting the center-to-center distance of the corrector vertex and the objective's optical axis. We do not want the corrector edge too close to the cone from the objective or clipping will occur. For this test, shoot for about a 2-inch clearance. See tube layout in Figure 11.8. Now adjust the test flat to bring the autocollimated beam to focus close beside the bulb. If you have a laser alignment tool this whole initial alignment will be much easier. With a low-power eyepiece examine the returned filament image for lateral color (dispersion). Adjust the field mirror in tip-tilt to remove (cancel) the color. Remove the bulb from its socket and replace with one with its surface fine stoned to a ground glass appearance. Make a rather large pinhole in aluminum foil and attach it to the bulb. Now examine the returned pinhole image with a ½-inch FL eyepiece. Tune out any remaining lateral color. By slowly traveling back and forth through focus you will see some astigmatism. If

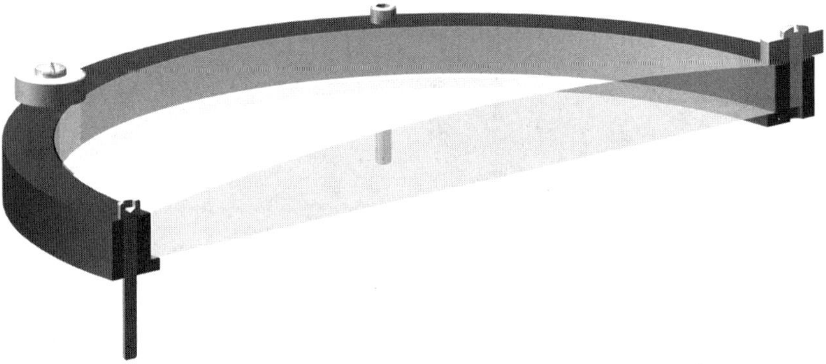

Fig. 11.3 *Sectional view of the objective in its cell. Drawing courtesy of Will Weidner.*

the astigmatic image flips horizontally and vertically then the objective probably only needs to be tipped. Tip the objective towards or away from the field mirror to minimize the astigmatism. Sometimes the objective is not properly square-on azimuthally, but by working each axis you will quickly converge to an astigmatism-free image. A helper makes all this much easier.

Remember that after each tilt iteration the field mirror will need tweaking, as you are not tilting the objective about the nodal point.

Next observe the color correction. By moving the corrector or the field mirror a few mm one way or the other along the optical axis, you will quickly find the color-corrected spacing. Of course these adjustments affect image position and lateral color slightly, so just keep touching these up. If your polishing and initial figuring were good you will see a rather nice image of that somewhat ragged but bright pinhole. The ground glass behind the pinhole sometimes provides little "star" images that, if you're experienced, can be used for additional tweaking. Replacing the "big" pinhole with a tiny one will allow you to see the beginnings of a diffraction image. Symmetry of the in- and out-of-focus images means you did really well! Usually, though, things will not be perfect, so on to the next test.

Replace the bulb assembly with your white-light knife-edge tester and cut the system off. If your work is done the surface will dim uniformly at focus, however, usually a slight zone is present and should be removed by figuring the front side of the corrector. The coated side is normally protected with black spray paint provided by the coating supplier (talk to them about this and any further protection they offer). In the test set-up the light traverses the first surface of the corrector four times, making the test sensitive to figuring action, so go easy.

Make sure the intermediate and final foci are laterally spaced correctly and that absolutely no trace of color or axial astigmatism is present before interpreting the shadows. Experiment with your system until you understand it thoroughly and are satisfied with the arrangement. Now, carefully measure the real spacing obtained so you can duplicate it in the tube.

The Schupmann telescope is unrivaled for the definition of fine lunar and planetary detail and the splitting of difficult doubles, yet is perhaps the most sadly neglected and most misunderstood instrument in the history of telescope making. The author hopes that this optical work section will open the door a little wider.

11.13 Reflective and Anti-Reflective Coatings

- **The Objective** It is recommended that the objective *not* be antireflection coated. The light gain is not worth the extra cost, light scatter (coronagraphs do not have coated objectives) and handling risk. It is so easy to occasionally remove and clean the uncoated objective with plain soap and water just the way we did during clean-up in polishing! This will keep the scatter low. One must distinguish between light scatter caused by small-scale surface roughness as opposed to surface reflections and ghost images, neither of which invade the medial's focal region in sufficient strength to be detected in the normal sense. Additionally, if CCD work is anticipated then any AR coating would need to be a very expensive broadband multi-layer.
- **Field Mirror** Overcoated aluminum has a reflectivity of about 91% and will provide years of trouble-free performance. Alternatively, protected silver jumps the reflectivity to 98% but has an unpredictable lifetime. A standard enhanced aluminum will easily exceed 94%. Some of the special dielectrics can attain 99% reflectivity, but probably not over the whole CCD bandwidth. Scatter could be a problem with some fancy coatings. The builder should opt for the lowest-scatter coating available that has over 91% reflectivity (see Section 10.3).
- **The Corrector** Silver is, by far, the best material for the reflecting surface. This is because of the high index interface at the back surface. Aluminum will provide 88%, but silver delivers 97.2% The silver coating will last for years as it can be protected with various paint layers, just like a shaving mirror. Also, as with the objective, no AR coating is advised on the refracting surface, although the objection is much tempered because in solar work this element is located beyond the occulter thus is not in full sunlight. The overall transmission of the standard medial including an enhanced fold mirror at the eyepiece (star diagonal) is about 72%, with 80% possible with special treatment of the field mirror. AR coating of the corrector could boost this to 88% or so.

11.14 The Tube, Cells and Stray Light Baffles

This section outlines a simple box tube designed specifically for the standard medial. The structure is illustrated in Figures 11.4 and 11.8. Other arrangements (see Chapter 3) will require structures tailored to their particular layouts. Regardless of

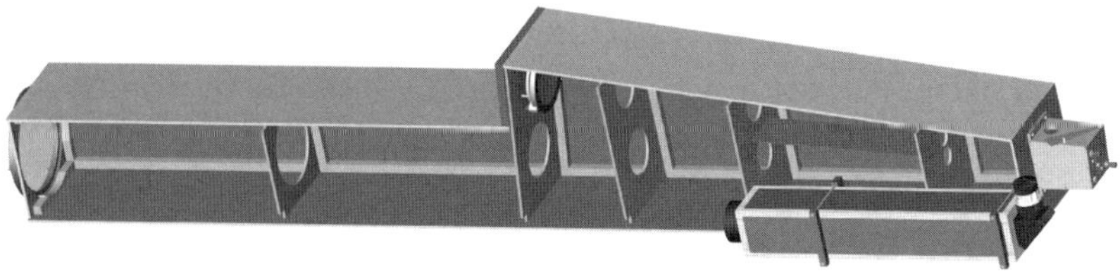

Fig. 11.4 *Cutaway view of the author's 9-inch instrument, a realization of the ideas shown in* ***Figure 11.8****. Note the separate field-mirror box in this example. The special sky baffle is also clearly seen. Drawing courtesy of Will Weidner.*

what optical arrangement you settle on, the principles are basically the same and the construction hints given here will apply equally to any medial. The wooden box-type tube has proven the most suitable structure to support and enclose the instrument's optics. The stiffness-to-weight ratio is hard to beat unless one goes to carbon fiber over-layers. The skin thickness can be ¼-inch marine grade plywood with mahogany the best choice all-around. The plywood is simply glued and tacked to a pine strip frame. Access areas to alignment adjustments utilize sailing dingy inspection ports suitably located near the corrector and field mirror. A port located close to the objective can help expel trapped hot air for early evening viewing. By pointing the scope straight up, opening the bottom and top port, a chimney is formed. Only use this trick advisedly, as dust and pollen are also swept into the tube! One side of the tube should be removable for large-scale access. Use lots of wood screws so stiffness doesn't suffer. For thermal boundary-layer avoidance make the box big enough to clear the beams by at least 1.5 inches at the objective, and 2 or more inches everywhere else.

The objective cell (preferably aluminum) need not follow the usual refractor construction in which a threaded spin-ring holds the crown and flint elements in close contact. This method leaves the first surface well below the plane of the cell rim and collects dirt and grit in the threads and in the corner of the spin ring which makes the objective's outer surface almost impossible to clean in a simple routine manner. The single element of the Schupmann can be mounted in a much simpler cell making occasional cleaning fast and sleek-free, especially important when using the instrument coronagraphically. This is made possible by machining the cell such that the thickened retaining rim is just flush with the objective's first surface. As the isometric drawing in Figure 11.3 shows, only three washers, secured with Allan head screws, are required to hold the lens in its cell. These nylon or Teflon washers are custom machined extra thick, say ¼-inch for a 10-inch lens. They are just large enough to safely catch the edge of the lens and still not form an optical obstruction. The washers are left round to prevent an accidental rotation leaving the lens free to fall out. Don't get cute here by making them cam-shaped to make lens removal easy! This design has no corners to collect dirt and grit, making it possible to make the final wipes of the lens (after being gently daubed and washed

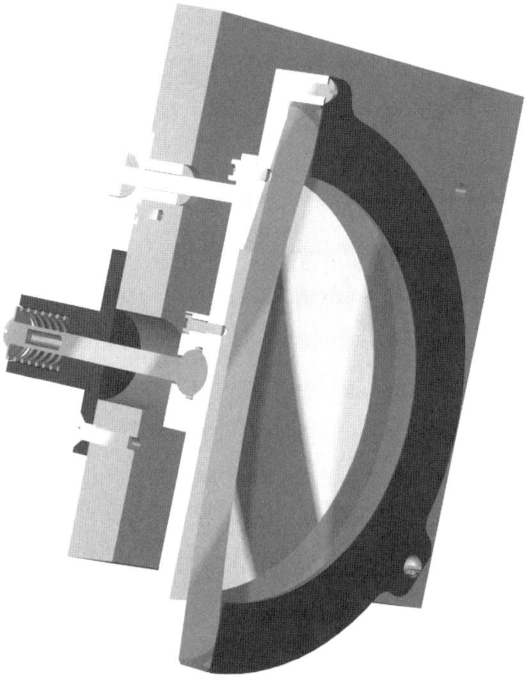

Fig. 11.5 *Cutaway view of corrector and cell for author's 9-inch medial. Here a spring pulls on the cell through a ball-joint against 3 tilt-adjustment screws. The corrector was originally made for a 10-inch instrument, thus is oversize. Note the Lyot stop which defines the instrument's clear aperture. See Figure 12.12 for entire instrument. Drawing courtesy Will Weidner.*

in the usual way) right over the cell edge. A little water finding its way between the lens and cell can be blotted away with a folded paper towel inserted between them. The remaining film soon evaporates (no spin ring to trap it).

First-class opticians will understand fully the reasoning for this cell design and will also know the proper, plain soap and water, method of cleaning dust and pollen off a fine coronagraphic lens without damage. Usually the scope is clamped aimed up slightly for cleaning the first surface. This allows the three retaining washers to be removed making cleaning the first surface really easy. Every few years enough particles find their way onto the back surface to warrant a cleaning, and a slight haze may have also formed there. Simply remove the cell with its lens and bring it to the kitchen. Place a clean, dishcloth covered, block of wood on the table. This block should be thicker than the cell by at least the lens thickness and wide enough to safely support the lens face up. Remove the retaining screws and Teflon washers. Now lower the lens cell over the block which leaves the lens on the block and the cell on the table. Clean the lens in a big plastic dishpan the same way you did on the original day of installation. Reinstall the objective lens by reversing the steps—pulling the cell up over the lens, lifting the lens by the cell and placing the works on the table and replacing the retaining screws and washers.

Cleaning is now complete.

The lens tilt is easily fine-tuned by shimming to remove the last vestige of axial astigmatism, the bulk of the tilt already built into the tube end. The author uses commercial color-coded plastic shim-stock and punches out disc-shaped "washers" nipped to resemble a horseshoe for easy slipping around the cell retaining bolts during this somewhat trial and error process. A very thin "O" ring can be lightly stretched over the cell to seal the shim gap.

The lens cell should be machined with about 0.020-inch (diameter) clearance, and plastic shims cemented in at 120-degree intervals. The shims should be of such thickness that some lens rattle is heard when the cell is shaken. The coefficient of expansion of aluminum and crown glass are quite different, so if you live where there are bitter cold nights be sure to compute the required extra lens clearance for your conditions. Alternatively, and still with adequate clearance, a third spring point could be incorporated in the cell structure. It would be designed to always apply a force equal to or slightly greater than the weight of the lens against the two fixed pads when pointing at the zenith. If, in your observing, you can avoid "pier flipping", the spring load can be lighter.

The rear supporting land should be bored to maximize the aperture, yet safely support the lens. A ⅛-inch land is fine for up to 10-inch aperture instruments. (Usually the working aperture is defined by the Lyot stop located over the corrector.) This land can have three evenly-spaced 1-inch areas of thin plastic tape to give some cushioning in supporting the lens. (In the Schupmann, objective lens deflection has no perceptible effect on image quality.)

Mount the cells on plywood end plates, perhaps ½-inch thick. This will ensure that the tube stiffness is carried through to the adjustment screws and shimming points. As mentioned before, the objective will need to be tilted about 0.6 degree depending on the corrector tilt. Knowledge of the *exact* tilt is not vital to predetermine at this point, however, you will get a pretty good idea of the tilt from the bench tests. Use the reflection of a laser beam off the back of the objective when the system is perfect on the bench, thereby getting the tilt angle. It is usual to build the tilt into the box as close as possible, then shim the cell in further bench tests of the completed OTA. Star tests may indicate a slightly different tilt due to stratification of the air that can occur in a horizontal bench test. This adjustment is permanent and readjusted only if the corrector tilt is reconfigured.

Mount the field mirror on a special kinematic tip-tilt base plate. The adjustments screws are readily accessible from the eyepiece location and project out the back of the tube. An example of a field mirror "cell" is shown in Figure 11.6. This adjustment is not permanent and requires occasional attention to tune out lateral color and to cancel atmospheric dispersion. Typically a medial will remain critically aligned for days depending, of course, on the tube stiffness and the workmanship on the cells and their attachment. Most tuning is to confirm a setting rather than improve it. The field mirror assembly will require a flip-in dust cover with a target dot or cross hair, so that when closed the target is directly over the mirror center. See Figure 10.3.

Fig. 11.6 *Field mirror and cell adjustment screws for the author's 9-inch medial. The thick Cer-Vit mirror is pressed firmly into the cell's land by a nylon screw on a flex stem. The adjustments are mutually perpendicular. Note kinematic push points, conical seat for fixed pivot point hidden from view. Diagonal mirror is at upper left.*

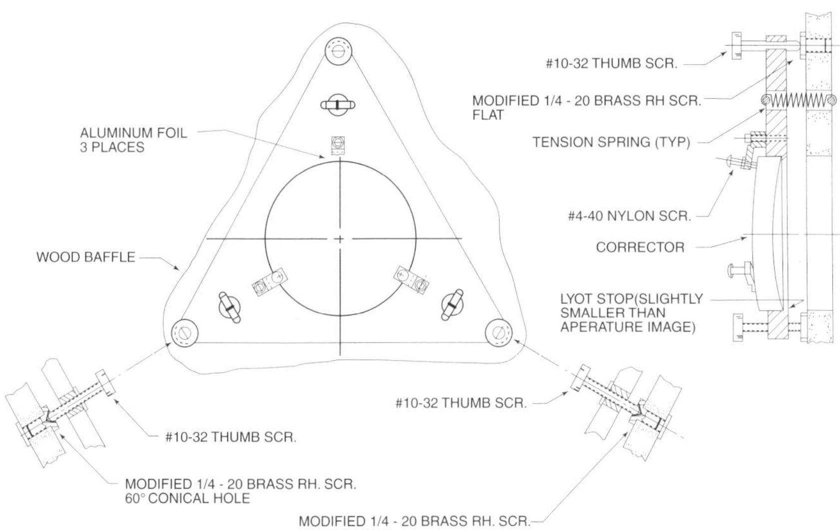

Fig. 11.7 *An example of a simple corrector cell design.*

The corrector cell will need tip-tilt and a small focus adjustment. Usually, the focus is done by advancing all three tip-tilt adjustment screws, equally. This focus adjustment is to remove the last trace of axial color if need be. A successful design is shown in Figures 11.5 and 11.7. This adjustment will remain stable for months, especially if you have protected the wood tube from moisture effects. The setting is quickly checked with the special alignment target at the field-mirror center.

Getting back to the box-like tube: although not shown in the sketch it is important to double the ply thickness in the region where the declination saddle will attach. Simply epoxy laminate the cut-to-size ply plates to the inside of the tube between the stray light stops. Carry the ply laminates right around inside the box. When mounting the tube on the saddle plate, be sure the declination axis is truly perpendicular to the optical axis. A line through the objective and field-mirror centers defines the optical axis of a medial.

Stray light stops (apertures in the bulkheads) are an important part of the standard medial design. There are two reasons for their inclusion in the tube structure. First, they provide great resistance to tube flexure by maintaining a rectangular and buckle-free outer skin. Simple stick ribs help but do not maintain a squared-up box-like structure. Second, the stops along with two special baffle plates, at least in the standard design, prevent the eyepiece from seeing the objective *directly*. Although this is unimportant for most planetary observations, it could prove to be a nuisance in lunar occultation observations. Here the field would be illuminated by off-axis skylight directly from the objective, adding to true field light. This would have the effect of reducing contrast and, in some rare cases, provide an uncorrected and rather colorful star image! For field lens configuration baffles, as such, are not necessary. This is because the system is inherently baffled or isolated by the field lens as this lens separates the objective section from the corrector section, and all unwanted stray rays fall on the flat-black Lyot stop (in solar work) and surrounding blackened structure.

The standard design, once stopped and baffled to prevent a direct view of the objective at the eyepiece port, makes use of the inherent dark field qualities gained by the action of the Lyot stop. The bulkhead "stops" are not true optical stops as they do not intrude into the beam. The Lyot stop, as mentioned earlier, is configured as part of the corrector retaining ring. This stop is a true optical stop and sets the working aperture.

The sketch of the tube in Figure 11.8 shows the recommended stops and baffle layout. When drawing the stray light stops/bulkheads in a full-size layout, use rays from the objective edge to the edge of the nominal 1.25-inch diameter field mirror, and again to the edge of the Lyot stop. Add about ¼-inch beam clearance for openings. It's a good idea to bore 1.5-inch diameter holes in the corners of the bulkheads for better air circulation. The space or clearance between the edge of the Lyot stop and the edge of the converging beam from the objective should be ⅓ of the clear aperture. This leads to a slightly greater field-mirror tilt than minimum clearance would indicate, however this set provides room for a very special baffle plate between the two beams. This plate and a smaller one labeled "sky baffles,"

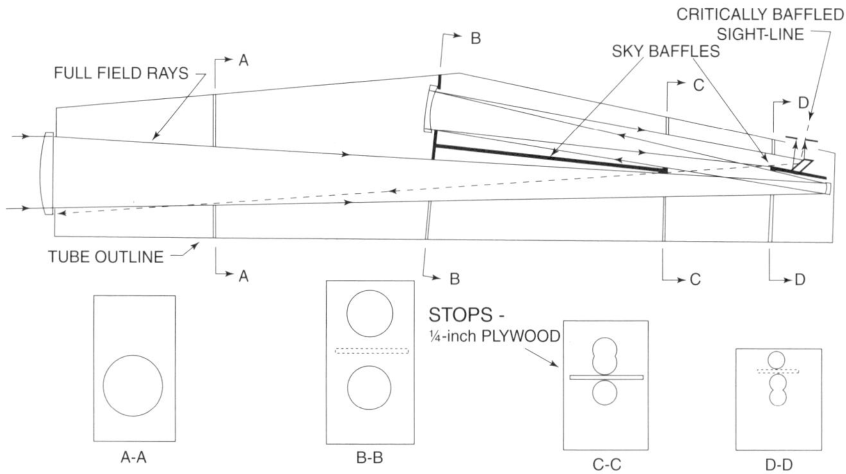

Fig. 11.8 *Plywood tube and baffles.*

are shown in the Figure. These two baffle plates are really effective in preventing a direct view of the objective. Since the plates are thin and inside the tube, their thermally induced boundary layer is of small scale and optically short. Make the plates narrow, thus giving room for air circulation. Thin aircraft ply (2mm) is perfect for these special baffle plates; however, the long edges should be stiffened with ⅜-inch square pine strips. Add light stops near the eyepiece as needed to prevent any tiny glimpse of the objective. All this work is done without the lenses in place; just the field mirror and diagonal tailpiece assembly needs to be opened up for checking. Use the screwed-on objective cell alone to define the aperture. It should go without saying that the entire inside must be painted or sprayed with high quality flat-black paint. The removable side makes this a fun job.

For the advanced ATM all this construction detail is probably unnecessary. The author hopes that enough details have been left for your own imagination, for one of the many joys of Amateur telescope making the invention of construction and test methods, and the challenge of pioneering unusual telescope optics.

11.15 Final Alignment and Star Tests

During the bench-testing phase the optical element spacing and tilts were adjusted to sufficient accuracy to carry out the final figuring work. If these settings have been faithfully reproduced in the final tube structure, the remaining adjustments will be easily accomplished. The adjustment steps follow:

- **Step 1 Aligning the Corrector.** Aim the tube at the northern daytime sky (far from the Sun!). Rotate or flip the alignment target over the field mirror. If you lack this feature, and you better incorporate it soon, simply provide a makeshift target centered on the field mirror. A black cross on white works well. Be sure the target and mirror centers are coincident to better than 1/16 inch. Tilt the corrector until the target can be seen through the empty draw-tube with your viewing position well back from the scope. Install a low-power eyepiece with a cross hair and focus on the target. The image will be very blurry as it is uncorrected for color and astigmatism. Tilt the corrector to center the cross and that step is done. Keep a special dedicated cross-hair eyepiece for future use to check the stability of the corrector alignment.

- **Step 2 Rough Adjustment of the Field Mirror.** Again in daylight, flip out the target and with the eyepiece removed stare down the drawtube (eye close now). Adjust the field-mirror tilt to center the objective over the corrector.
- **Step 3 Refining the Field Mirror Tilt.** Under a night sky find a ~4th magnitude, high elevation star, and center it in a low-power eyepiece. The star will show a short spectrum, red to violet. Adjust (tilt) the field mirror first in an axis that minimizes the spectrum length, then in the other axis to nearly eliminate the spectrum. Install a high-power eyepiece, say 150 diameters, and reduce the spectrum to zero.
- **Step 4 Correcting Axial Astigmatism.** Rack through focus and observe the residual astigmatism. Perhaps you got the objective tilt just right and there will be none! The best way to evaluate astigmatism is to observe the image somewhat away from focus where it shows an oval shape. Tilt the objective slightly and note whether the image is becoming rounder. You will quickly determine the correct tilt axis with respect to the major axis of the oval extra-focal image. As you hone this adjustment work closer to focus. Each time the objective is adjusted you may have to touch-up the field-mirror pointing. The author does the objective tilting with plastic shims cut in a horseshoe shape to slide on either side of the cell's hold-down screws. This adjustment, as mentioned before, is semi-permanent.
- **Step 5 Tuning Out Axial Color.** Provision to move the corrector along the axis a few mm was mentioned earlier. This spacing adjustment is utilized to remove any residual axial color. If you have mounted the optics such that the spacing is within 1mm of the test bench optimization, then no color residual will be seen. However, the usual case is that slight color will be seen when the image is de-focused enough to see many rings. The color residual will show best in the center of the pattern. If inside focus a purplish red is observed, simply move the corrector one or two millimeters toward the field mirror. If that color is observed outside focus go the other way. To do these adjustments it's a good idea to mark the corrector adjustment screw heads with an arrow so a one-turn advance is easily judged. Always fine-tune the corrector tilt with the field-mirror target, as well as the lateral color after each axial color adjustment. This should end the adjustment phase. The image of a star will be absolutely color-free and perfectly round with one or two diffraction rings, com-

pletely unbroken, surrounding the central disk. If the first or second ring is broken in 4 places and the breaks remain stationary in position angle than you may have residual astigmatism. To be sure though, observe an overhead star to rule out unsymmetrical air stratification effects in the tube. This is especially true when the system has not reached equilibrium with the night air. Of course, all of this fine-tuning requires good seeing.

Point the Schupmann at the Moon or a planet. If the seeing is really steady be prepared for image sharpness and contrast that are emblematic of truly diffraction-limited optics!

11.16 Spot Diagrams

Every ATM seems to like spot diagrams, and why not—it's like looking into a magical eyepiece that, along with the telescope, ignores the wave nature of light. A computer-generated bundle or symmetric array of parallel pencils, usually hundreds, enters the optical system and is refracted or reflected at each surface in a mathematically perfect way. These bundles can be "sent" parallel to the optical axis, or angled to it. We see in the computer display the intersection of these ray bundles in the focal plane for any number of field points we desire. Each field point has a corresponding object point usually located at infinity. It's just a matter of selecting various object input angles to generate spot diagrams representing the geometric performance for those angles in the focal plane.

Ray trace programs allow us to input "rays" of different monochromatic wavelengths. Usually three, well separated, wavelengths are enough to get a good idea of the telescope's performance in a more or less quantitative way, especially if the color curve is handy. Analysis colors are chosen to best characterize the instrument consistent with its final use. For instance, visual use demands good energy concentration over the spectral region *C* (red line at 656.3nm) to *F* (blue line at 486.1nm). These are Fraunhofer lines of hydrogen. An in-between wavelength, say, the yellow helium "*d*" line at 587.6nm is pretty much standard for completing the 3-wavelength set for evaluating a visual refracting telescope's performance. This is a somewhat lax criterion and unless the instrument degrades gracefully at each end of this spectral span, the visual performance will be slightly compromised. Generally, the best performance is achieved when the *C* and *F* lines come to a common focus. CCDs have a much broader spectral sensitivity range than the eye, and designers strive for good energy concentration over the region 360 through 900 nm.

The spot diagrams presented here are selected from various designs tabulated in this book. Most medial systems are tilted-component designs and present asymmetric image shape about the field center; whereas untilted systems, such as the brachymedial, show symmetric image shape about the field center. This means that, for the untilted designs, we need only explore the field along one radial direction. However, with tilted systems we must examine the geometric image structure for at least four points, disposed at 90-degree intervals, around the field

Table 11.1
Tilt Angles in Degrees

Instrument Spots	Objective	Field Mirror	Corrector
Figure 11.9 8-inch *f*/12.5	0.63	3.6	0.60
Figure 11.10 4-inch *f*/12.0	0.85	3.0	0.60
Figure 11.11 6-inch *f*/15.0	0.50	3.0	0.35

edge, as well as the field center. For axially-symmetric spot diagram presentation it is usual to show the center, 70% and edge field points as presented here. For the tilted systems the author prefers to show the field center, the rim of the diffraction-limited "scientific" field and the full field. The full field often fails in terms of diffraction-limited performance, but is almost always just fine for wide-angle and necessarily low-power observing.

Each spot is shown centered in a circle, which represents the Airy disk diameter. Spots that show most of the energy well within this circle, say ½ diameter, are usually associated with diffraction-limited performance; however, the interpretation of spot diagrams can be tricky and only the MTF (modulation transfer function described in Section 4.1 on page 47) or Strehl ratio should be used for exact analysis. The Strehl ratio is a "goodness factor" with 1 representing perfection—that is, all the energy theoretically possible, consistent with the wave nature of light, is contained within the Airy disk. A Strehl ratio of 0.9 is considered excellent for most scientific applications, indicating that 90% of the possible energy is within the Airy disk. The Airy disk circle is the diameter to the first minima and one must consider, even for a perfect system, the non-trivial energy that lies outside of the disk in the form of diffraction rings. Usually, as the Strehl ratio degrades, the lost energy is deposited into the rings, thus lessening contrast.

Each spot diagram caption contains the instrument name, Table reference, objective diameter, *f*-ratio, field points in degrees off-axis, wavelengths traced, Petzval curvature and in some cases a second trace on a tilted but flat field is also shown. The rationale for presenting all the spots in one chapter, rather than along with each prescription in turn, is so the reader may quickly review the designs. But be warned—nothing is held constant, *f*-numbers and apertures cover a wide range. Spot diagrams are only a part of the instrument's "voice" and one must look at all factors that are detailed for each telescope presented throughout this book. Each design is set up to show its best face consistent with practicality as, in the author's opinion, direct comparisons (same aperture and focal length) of these varied and complex designs would be unfair and surely miss the point. Each design is specific and, for the most part, uncompromised.

The first 3 medials are tilted-component systems. Table 11.1 gives the individual element tilt-angle values in degrees. These data may be helpful to the builder in bench testing and laying out the tube structure. Remember that the tilt of the objective is orthogonal to both the field mirror and corrector for in-plane medial designs.

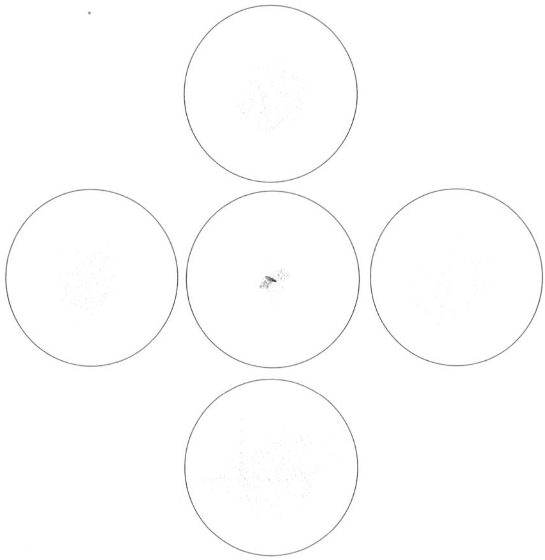

Fig. 11.9 *8-inch f/12.5 All-Spherical Super Schupmann images shown on a tilted, but flat focal plane. Wavelengths traced are 404, 580, and 1,014 nm. Field tilt: X = 1°18′, Y = 1°18′ FOV = 0.2° Petzval radius = 9.9 inches. See Table 5.1 on page 58.*

Medial Spot Diagrams:

1. 8-inch *f*/12.5 All-Spherical Super Schupmann (see Figure 11.9)
2. 4-inch *f*/12.0 Introductory Medial (see Figure 11.10)
3. 6-inch *f*/15.0 Quartz Medial (see Figure 11.11)
4. 8-inch *f*/12.4 Willard All-Spherical Split-Field Anastigmat (see Figure 11.12)

Brachymedial Spot Diagrams (Note: All brachymedials are traced without diagonal mirror obstructions. The choice of diagonal diameter is too great to include various subtle effects, but keep in mind the MTF curves and spots will be modified, more so with relatively larger obstructions):

1. 6.8-inch *f*/23.5 Type A All-Spherical (see Figure 11.13)
2. 6.8-inch *f*/27.9 Type B All-Spherical (see Figure 11.14)
3. 9.1-inch *f*/27.2 Special Type B (see Figure 11.15)
4. 8-inch *f*/7.4 Honders Planetary Brachy (see Figure 11.16)
5. 6-inch *f*/2.7 Honders Brachy Camera (see Figure 11.17)

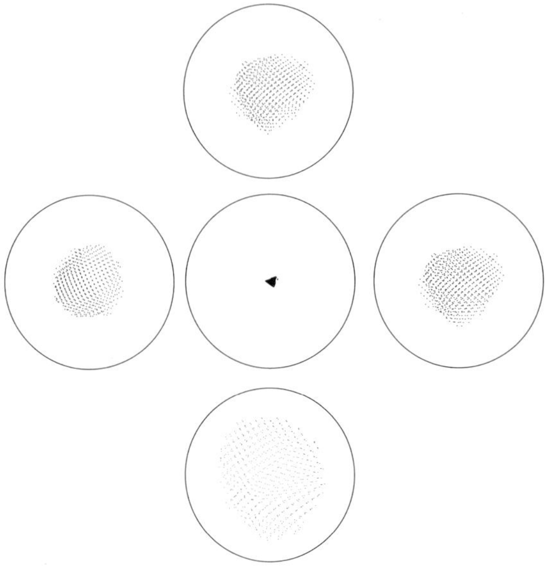

Fig. 11.10 *4-inch f/12 Introductory Medial: Images shown on a tilted, but flat focal plane. Wavelengths traced are 486, 587, and 656 nm. Field Tilt: X = 1°20′ , Y = 0°24′ FOV = 0.4° Petzval radius = 9.7 inches. See Table 5.2 on page 60.*

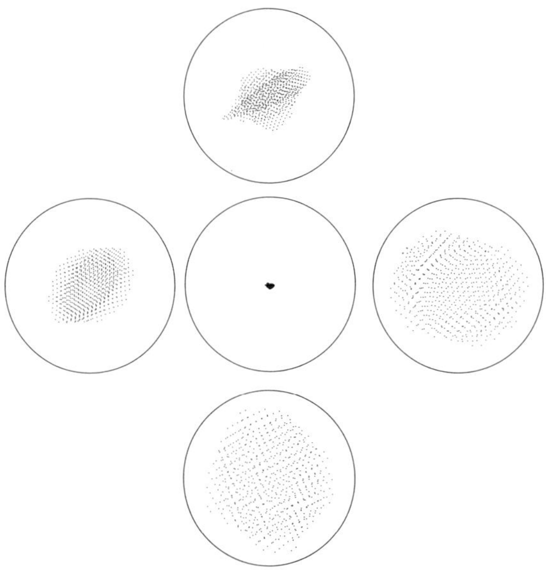

Fig. 11.11 *6.3-inch f/15 Quartz Medial: Images shown on a tilted, but flat focal plane. Wavelengths traced are 436, 587, and 852 nm. Field tilt: X = 0°51′ , Y = 0°10′ FOV = 0.5° Petzval radius = 18.1 inches. See Table 5.3 on page 63.*

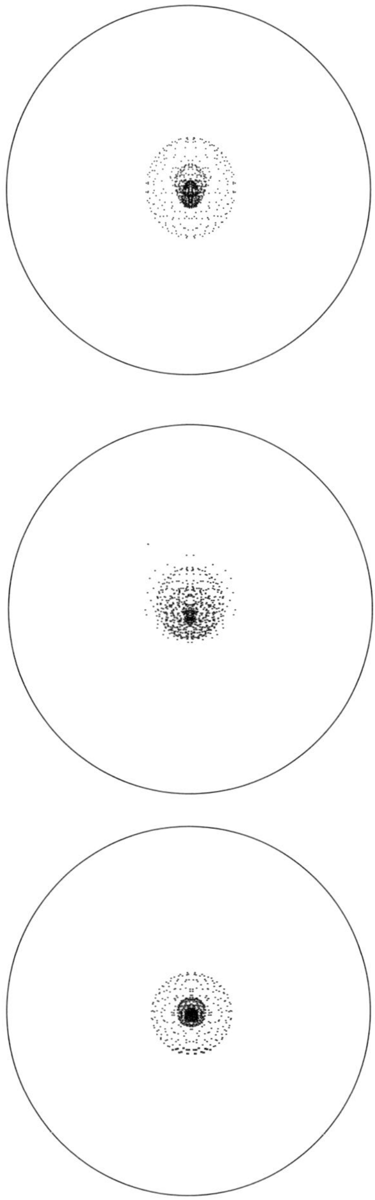

Fig. 11.12 *8-inch f/12.4 Willard All-Spherical Split-Field Anastigmat Medial. Image presented on a tilted best-fit off-axis flat surface. This radial tilt is 1.6° in this example. Wavelengths traced are 486, 587, and 656 nm. FOV = 0.23°, Petzval radius = 16.9 inches. See Table 6.2 on page 71. The spots are arranged as follows: Lower spot is 0.069° from optical axis and represents the field edge next to the pick-off mirror. The center one is the working field center located 0.185°off-axis. The top spot is for the field edge and is 0.30° off-axis.*

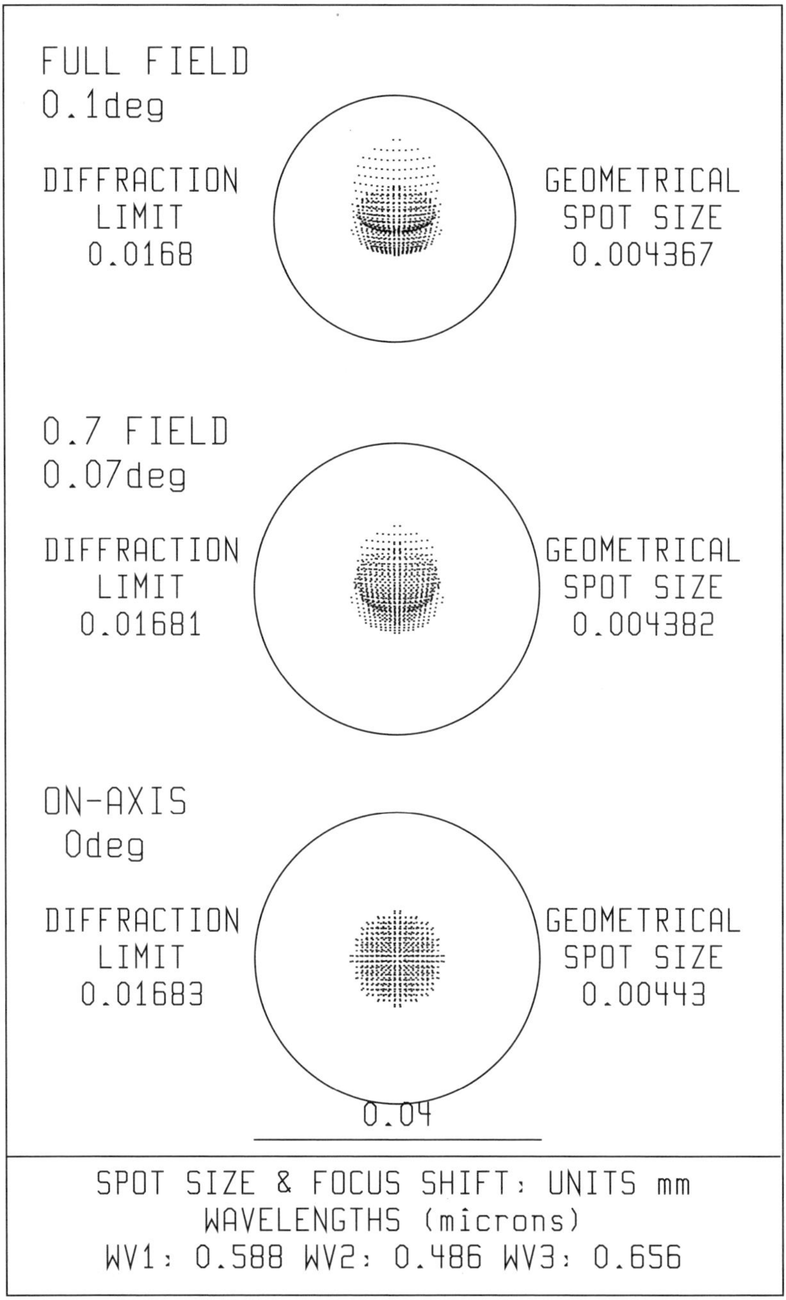

Fig. 11.13 *6.85-inch f/23.5 Type A All-Spherical Brachymedial. Best-focus flat focal plane. Wavelengths traced are 486, 588 and 656 nm. FOV = 0.2°, Petzval radius = 51 inches.* *See Table 9.1 on page 99.*

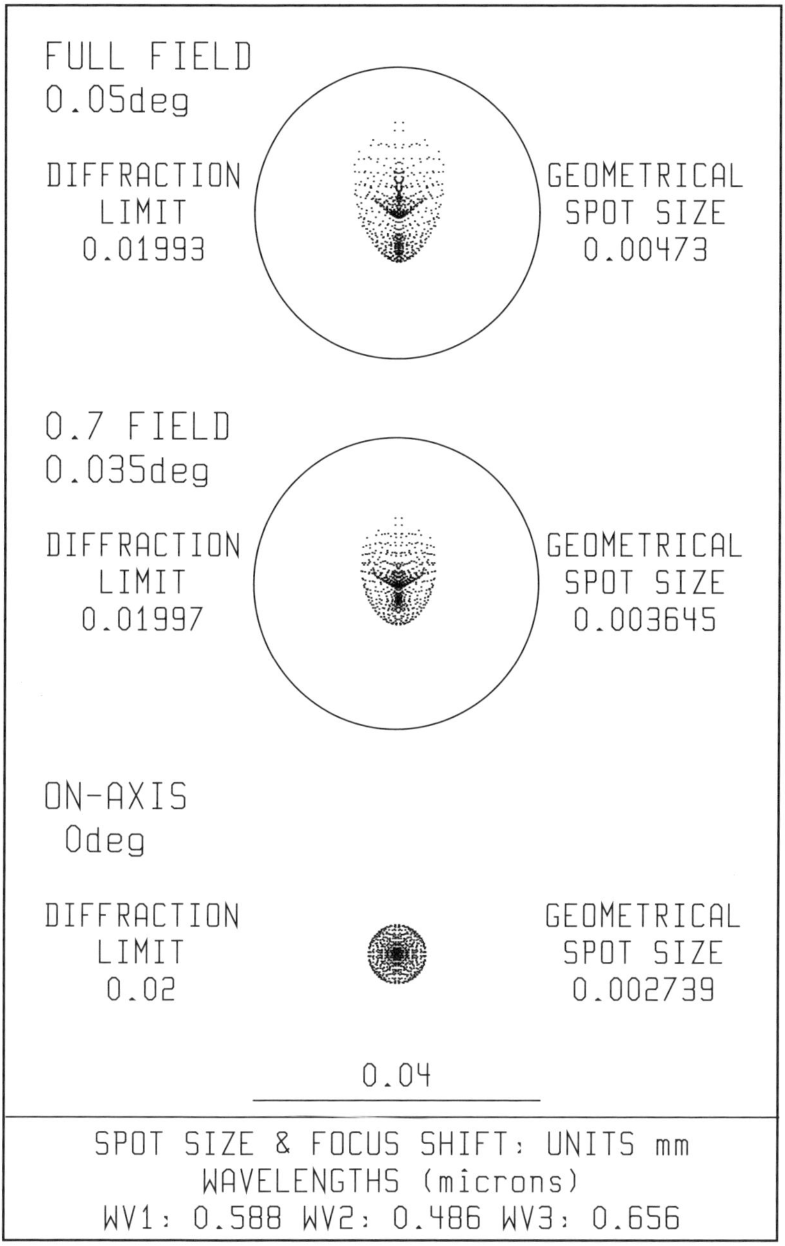

Fig. 11.14 *6.8-inch f/27.9 Type B All-Spherical Brachymedial. Best-focus flat plane.Wavelengths traced are 4.86, 588 and 656nm. FOV = 0.1°, Petzval radius = 16.5 inches. See Table 9.2 on page 102.*

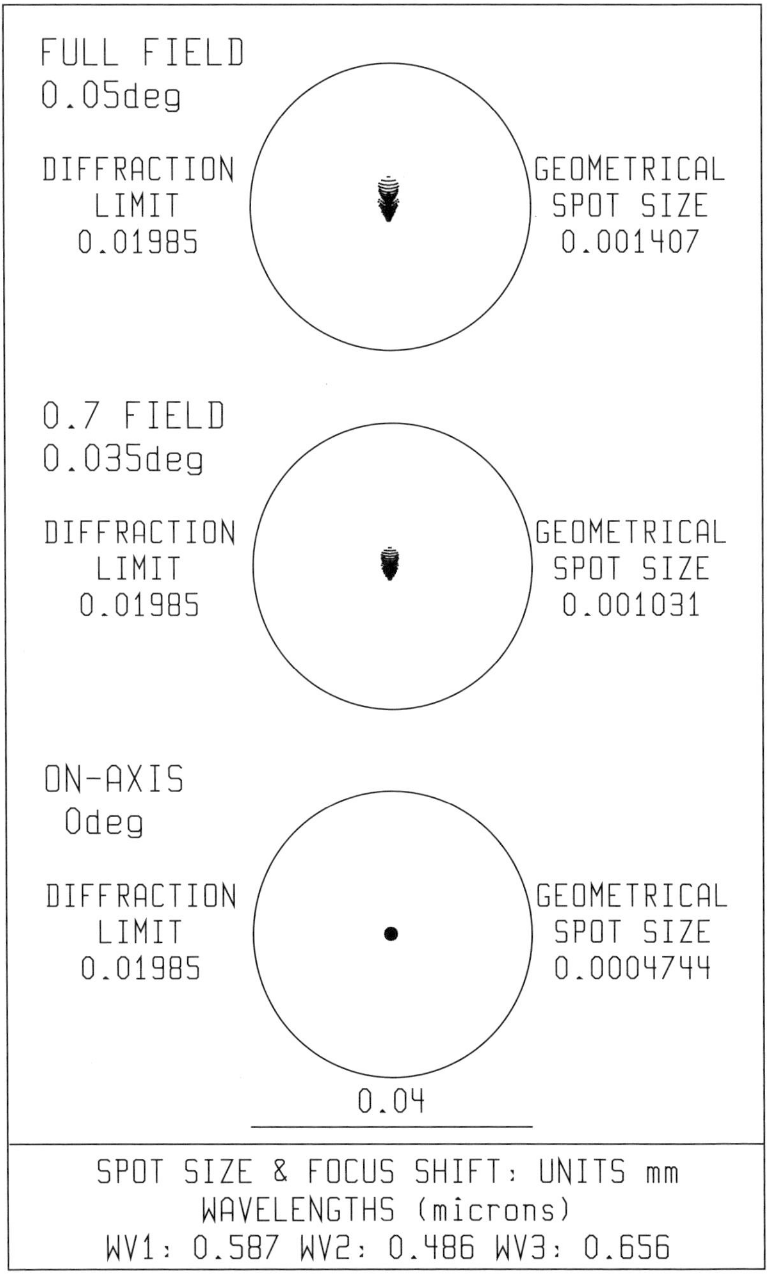

Fig. 11.15 *9.1-inch f/17.2 Special Type A. Best-focus flat plane. Wavelengths traced are 486, 587 and 656 nm. FOV = 0.1°, Petzval radius = 51.3 inches.* *See Table 9.3 on page 104.*

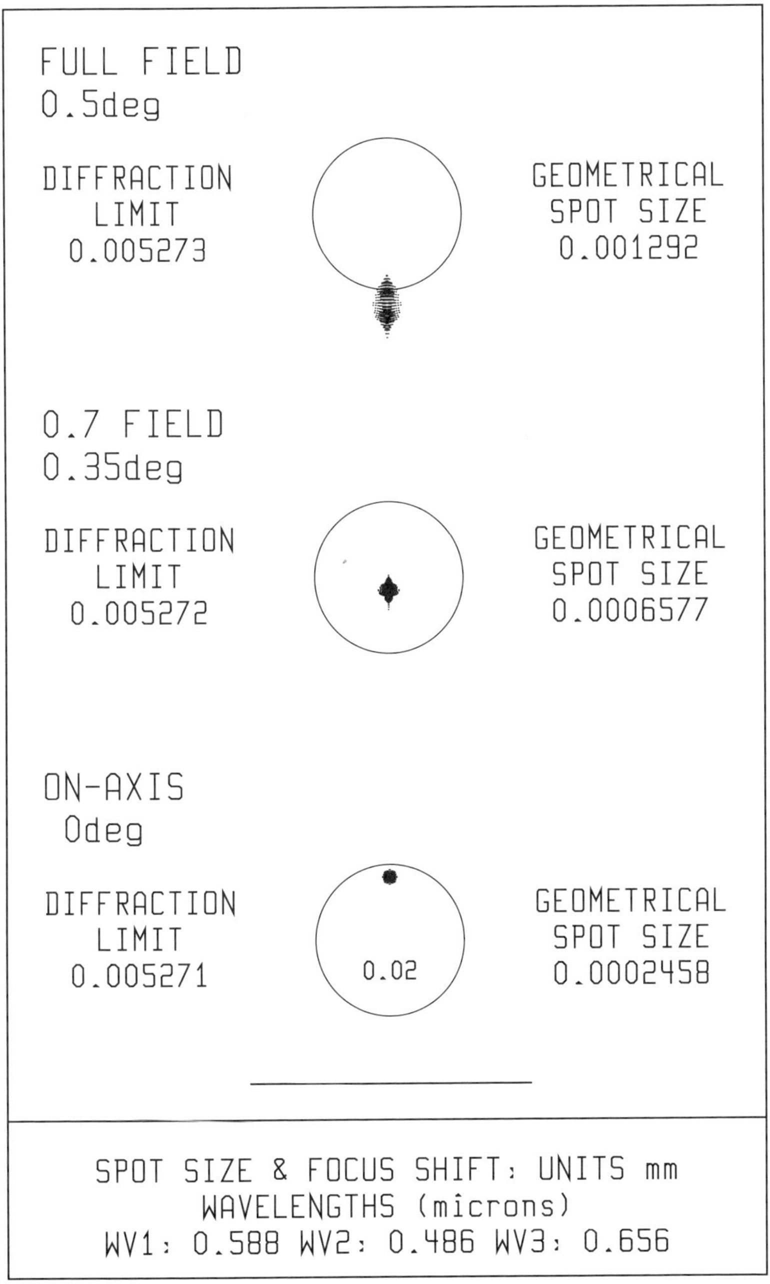

Fig. 11.16 *8-inch f/7.4 Honders Planetary Brachymedial. Wavelengths traced are 486, 588 and 656 nm. FOV = 1.0°, Petzval radius = 105.8 inches. See Table 9.4 on page 105.*

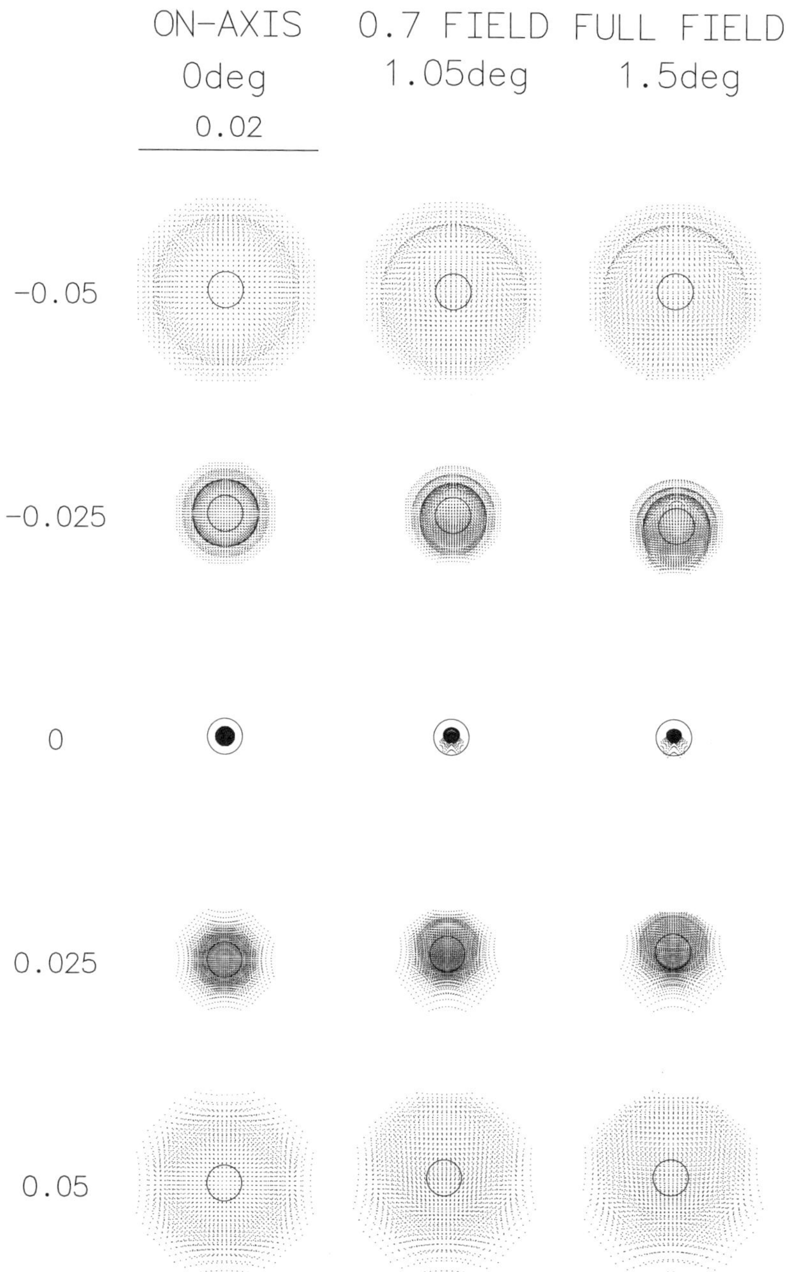

Fig. 11.17 *6-inch f/2.7 Honders Astrographic Camera. Wavelengths traced are 485, 587 and 656 nm. FOV = 3.0°, Petzval radius = 318.1 inches. See Table 9.5 on page 107.*

Chapter 12
Photo Gallery

12.1 Springfield Telescope Makers' 13-inch Medial

Located in the McGregor Observatory of the Springfield Telescope Makers, this medial was, until the recent completion of the 1-meter in the Canary Islands, the world's largest (see Figure 1.10). The observatory and the computer directed alt-az mount portions of the project were headed up by club member, John Martin. Scott Milligan with Philip Rounseville as principal optician, directed construction of the Schupmann itself. The plaque on the pier recognizes the many club members and others who made the observatory's successful completion possible. The instrument follows the specifications (scaled up) of the standard all-spherical design detailed in Chapter 5.

Fig. 12.1 *The Springfield Telescope Makers' 13-inch f/10 Schupmann medial mounted in the McGregor Observatory. Note the fine mural by club member Russ Camilla.*

Fig. 12.2 *The author is seen here (1984) with the first Super Schupmann.*

Because the mount is not an equatorial, certain scientific projects such as CCD binary star position-angle measures are made difficult due to field rotation. Although corrections derived from the pointing commands can be applied to the position angle relative to the CCD frame, plans are being made to incorporate a high precision computer-controlled mechanical field de-rotator. This will be especially useful in long-exposure CCD photography. Currently, astronomical work is centered on CCD photometry of faint eclipsing variables and binary stars.

12.2 Daley All-Spherical Medial

The author, seen in Figure 12.2, constructed this 6-inch *f*/15 medial in 1984 to test the practicality of the all-spherical standard design described in this book. The instrument is the world's first Super Schupmann and is distinguished from Kutter's all-spherical medial (see Figure 3.8) in that the color correction is perfect, whereas Kutter employed light flint in the corrector leaving a small color residual of about half that of the best apochromat of that era.

Years ago this instrument was conveyed to the Boston Amateur Telescope Makers where it is presently being used for planetary observing. It is shown here in the author's backyard observatory. The mount itself is an old Mogey equatorial head designed to carry a 4-inch refractor. Note the use of a European-type tube brace, and a 4-jaw lathe chuck as a counterweight.

12.3 Daley Solar Medial

While observing (without the Lyot stop) with Olson's medial, (Figure 12.8), the author constructed this 7.5-inch *f*/13.3 in-line field-lens type medial and in 1969, after returning Olson's instrument, mounted it on the "Porter Folly" mount in the backyard. You see, in those days we simply made our scopes absolutely weather-proof and left them outside ready to use—and what wonderful images they produced. Look close for the wire that led to a 50-watt light bulb mounted inside the scope near the corrector. This prevented all internal "dewing" and was disconnected about an hour before viewing. The late Chester Cook of the Boston Telescope Makers inspired this instrument, in part. He made a similar 8-inch field-lens system and it too was mounted in the open air. Although the author's instrument gave great lunar and planetary images, it was the desire for high performance in solar prominence observations that drove the development of this design.

This developmental path is a natural offshoot of the Schupmann's coronagraph-like optical layout. Lyot stopped to 7.0-inch, the instrument was used for many years to view prominences using a wire-supported occulting disk, which could be flipped in place over the center of the field lens. This simple feature allowed the continued use of the instrument on the Moon and planets with minimum fuss. Of course, the usual Lyot stop (also flippable) was positioned over the corrector, without which the (prominence work) contrast would be seriously compromised (see Chapter 7). Although a narrow band (~100Å) H interference filter was normally used, it was fun to demonstrate the instrument's freedom from scattered light by viewing the prominences using a simple long-pass red filter, such as a Kodak #92, mounted in the eyepiece. The simple filter worked best under "coronal skies" where it competed favorably with the interference filter particularly in terms of image brightness. Readers should be warned that a really solid mount and accurate drive are required when using wide-band red filters in this type solar of instrument, as even the limb is mighty bright should it peek out from behind the occulter!

12.4 Anton Kutter's Third and Last Medial

These 1946 photographs in Figure 12.4 show the nearly completed all-spherical 5.9-inch *f*/11.3 "Else Kutter Medial." Although the objective lens is of the usual crown glass, the 1.58-inch diameter corrector is of light flint, necessary for a spherical solution. Even with this mismatch in glass types the color correction is a factor of ten better than a standard *f*/15 achromat. Kutter, in a private communi-

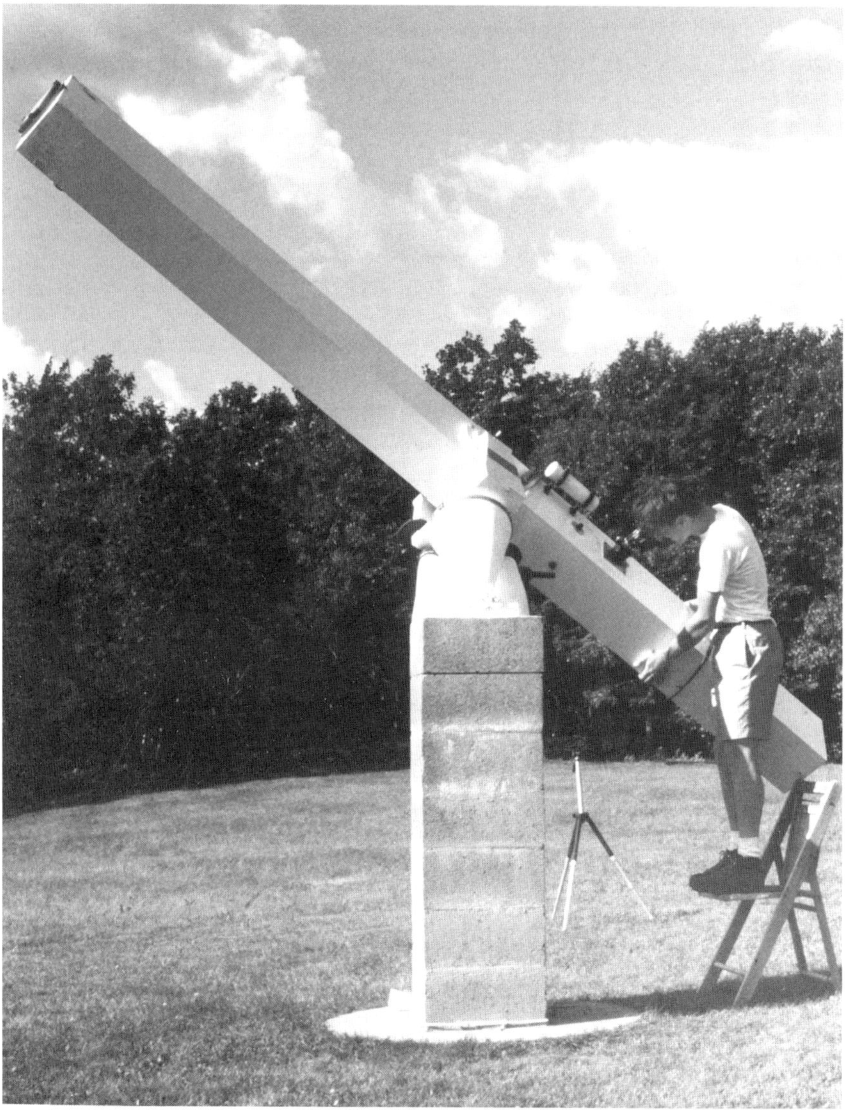

Fig. 12.3 *The author's 7.5-inch solar/planetary medial mounted in the open on a "Porter Folly" mount.*

cation with the author, ended the technical discussion of the instrument's chromatism with these words: "This color correction overtops that of a Zeiss-half-apochromatic OG of *f*/17.5 with a factor of 2. Indeed, at a power of 330 times no trace of color is visible. All optical surfaces are spherical".

Kutter made good use of the tube with his skewed optical path, an idea seen in Schupmann's patent in Chapter 1 where the converging rays from objective run

Fig. 12.4 *The "Else Kutter Medial" is shown on the left with the cover for the corrector assembly in place. The right-hand view shows the engraved tailpiece assembly; note the two small adjustment knobs to center the entrance pupil on the corrector.*

obliquely to the back edge of the tube. Contained inside the engraved back plate is a field lens and just a short way further down the path, a small optical flat which directs the now-diverging cone at a 40-degree angle through the side of the tube to the corrector element as detailed in Section 3.2 and Figure 3.8. The corrector is tilted in the usual way to return the beam to focus at the eyepiece assembly just beyond the field elements. A clever layout indeed! This instrument was the first to employ a single-element corrector, and is of tremendous historical interest as it set the path for all future amateur interpretations of the Schupmann principle. Even today one would not be disappointed with this design, and for those who are interested in exploring the small corrector solution, it is still a good place to start.

12.5 Larry Prugnarola's Medial

The late Larry Prugnarola loved to observe the Sun and Moon with his 7½-inch *f*/13.3 medial. He designed a special field-mirror cell to facilitate interchanging different versions of this component. For prominence work he made a perforated field-mirror with the hole just a little bigger than the diameter of the image of winter Sun, letting the light and heat go out the back of the scope. You did need to take care to keep your shirt and tie out of the emerging beam!

The limb of the sun was brought to the edge of the hole and by use of a red

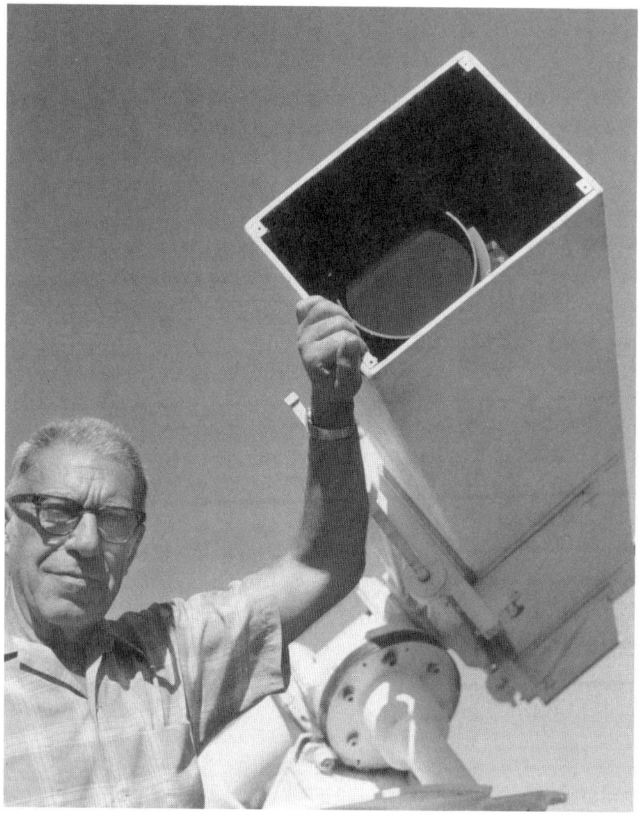

Fig. 12.5 *Larry Prugnarola is seen here with his beautifully crafted 7.5-inch f/13.3 medial. The corrector is ½ the objective's diameter. The instrument is carried on a very rugged equatorial mount that is thought to have been made by the late Frank Noyes of the Boston ATMs.*

eyepiece filter the prominences were beautiful to see. The solar limb was scanned by an eccentrically located, rotatable eyepiece holder, so that even a high power ocular showed the prominence centered in the field. Other field mirrors included an un-coated one for white-light solar observations and an aluminized one for lunar and planetary work. Each mirror was mounted in its own sub-cell for safe handling (see Chapter 7).

12.6 Gerry Logan's Folded Medial

Gerry Logan of Lakewood, CA is seen here at the eyepiece end of his beautiful 10.25-inch *f*/9.75 Schupmann. To the author's knowledge this is the world's first folded medial. By folding with a 6-inch flat, the optical path length is only a little over 4.5-feet, making the instrument quite portable, allowing Gerry to observe under dark skies. A lightweight black cloth shield surrounds the rods when observing. In addition to preventing dew and stray light problems the shield may also help to

Fig. 12.6 *Gerry Logan can be justly proud of his unique folded 10.25-inch f/9.75. The instrument is used primarily in the observation of faint planetaries. The black cloth light shield is removed here for clarity.*

keep body heat away from the aperture. The instrument is primarily used to observe tiny, star like, planetaries. By adjusting one of the field mirror tilt control knobs, seen located skyward of the focuser, the planetaries are easily identified as the stars all show a short spectra, whereas the planetary remains sharp due to the dominant green oxygen line. Mr. Logan writes, "The performance of the 10¼" is awesome. To use it to find extremely small planetaries is incredible".

This is Gerry's second medial. His first, a 7-inch *f*/10 is an all spherical Super Schupmann, and is also based on the standard design described in this book. These two Schupmann's have received merit awards at the Riverside Telescope Makers Conference.

12.7 Mike Mattei's Medial

It was in the late 60s when Mike Mattei, an avid solar system observer, viewed the Moon through the author's 7.5-inch *f*/13.3 medial (see Figure 12.3). Mike never forgot the incredible contrast and sharpness of tiny lunar features seen that night. It was many years later that he found the time to build his own Schupmann. He chose a 2.3-inch diameter corrector for his 6-inch *f*/15 planetary scope. This instrument employs a field mirror as in the standard design. Due to the small corrector, a fair amount of aspheric work was required. It was felt that the larger objective would respond better to achieving a smooth aspheric so the corrector's surfaces were figured spherical and the back surface silvered thus completing this element. The tube and cells were soon finished and the optics in place.

Fig. 12.7 *Mike Mattei and his 6-inch f/15 medial. Visual observation of the planets, especially Mars, has been his lifelong passion.*

Fig. 12.8 *Edwin Olson's 6-inch f/15 medial mounted for testing on the author's "Porter Folly" fork mount in 1966.*

This facilitated the final work on the objective, as it was set up to be easily removed for spells of figuring. A beautiful smooth cut-off under autocollimation was achieved just days before the 1983 Stellafane convention where Mike won first place in the optical competition with his fine new medial. He has enjoyed making observations with it, both visual and with his CCD, up to the present day. Mike is seen here in 1988 with his equatorially mounted Schupmann telescope. Note the plastic sailboat deck port to access the corrector for adjustment and the telephone-pole pier.

12.8 Edwin Olson's Medial

The 6-inch *f*/15 Schupmann pictured in Figure 12.8 was designed by Edwin Olson and completed by him in 1964. This instrument represents the first truly successful interpretation of the Schupmann principle in the United States. The telescope design is pioneering as it was the first medial to employ a "full size" corrector (the same diameter as the objective) as well as the first to utilize a field mirror and is thus of great historical significance. This form has many theoretical advantages such as the reduction of (normally tiny) spherochromatism to essentially zero, very small compensating tilts, as well as low curvature of field and field tilt. The corrector by itself is corrected for spherical with spherical surfaces, leaving only the objective to aspherize.

This instrument is the first to employ Ed's idea of an out-of-plane optical layout, which has some interesting attributes.

In practice this form of medial is limited to about 9-inch aperture due to thermal effects encountered in figuring the corrector to a good sphere. The instrument is seen here in 1966 temporarily mounted for star testing on the author's "Porter Folly" mount. A good view of the tube baffles can be seen in the October 1964 issue of *Sky and Telescope*, pages 204–205. For its aperture, no Schupmann has outperformed the "Olson Medial".

12.9 Bert Willard's Solar Medial

This 4¼-inch *f*/11.3 instrument, nearly completed by Edwin Olson, was finished by the author and presented to Bert Willard in recognition of his many valuable contributions to the development of the Schupmann for amateur construction. To the author's knowledge, this instrument is the only (currently) operating medial utilizing a "full size" corrector. Bert has modified it for the observation of solar prominences in Hα light. Rather than a "occulting hole" in the field mirror, Bert has cemented an aluminum occulting cone to the field mirror face with its sharp-edged rim nearly touching the mirror surface which is, of course, positioned at the objective's Hα focus. The cone face is polished and in use appears black in the eyepiece even though the corrector is "looking" directly at it (see Chapter 7). Light reflected from the cone is simply absorbed by the flat black tube walls with no apparent thermal problem. Bert's method of scanning around the limb of the Sun is totally unique. Rather than the standard eccentrically-mounted eyepiece, the corrector is tilted to center the prominences. This method has proved very successful and is without visually detectable image degradation.

Looking at Figure 12.9 you can see the rotatable tilt control rods leading from the eyepiece region up to the top of the tube where fine thread screws adjust the corrector tilt. A wide band (100 Å) Hα interference filter, which is tuned by tilting, is located in the eyepiece assembly. A flat-black Lyot stop is located in near contact with the corrector and intercepts the bright, sharply imaged, diffracted light from the rim of the objective which greatly reduces the background glow allowing nice views of the prominences.

12.10 Gerry Logan's 7-inch Medial

Gerry Logan constructed this 7-inch *f*/10 medial to the standard all-spherical design called the Super Schupmann presented in the author's 1984 Schupmann con-

Fig. 12.9 *Bert Willard's 4.25-inch f/11.3 solar prominence medial. Note the handy control rods to tilt the corrector for prominence-centering in a high-power eyepiece. This instrument rides piggyback on a 10-inch Newtonian, which is used with a Baader filter to observe the photosphere in white light.*

Fig. 12.10 *Gerry Logan's 7-inch f/10 medial at Riverside.*

struction booklet, and this was probably the first medial that resulted from that writing effort. Gerry finished it in 1990 and displayed it at the Riverside convention where it got much attention and the builder, a merit award.

This instrument is the first US medial to be built with a focal ratio as fast as *f*/10, which is as fast as recommended. Gerry went on to top this by building an even faster 10.25-inch *f*/9.75 version. That instrument is seen in Figure 12.7. There is no harm in operating at these *f-numbers* for narrow scientific fields of view; however, the sensitivity to adjustment to correct the field center does increase inversely with the square of the focal ratio. Details for the standard design and its variations can be found in Chapter 5.

12.11 Jim Daley's 8.3-inch *f*/12 Split-Field Schupmann

This instrument, built by the author in 1990, was used primarily in the measurement

Fig. 12.11 *The author and his 8.3-inch (effective) f/12 split-field Schupmann telescope. The insulation on the concrete pier reduces daytime heating of its large mass, thus reducing nighttime local turbulence.*

of double stars with a filar micrometer and later by CCD. It represents the first known example of a split-field medial. The original plans for critical operation with its full 9.4-inch aperture failed. However, by (Lyot) stopping the aperture to 8.3 inches the performance was up to the task. At the full 9.4-inch aperture, knife-edge tests showed a fine cut-off of the objective and corrector in Hg green (monochromatic) light (they were separately null figured), but upon star-testing, the instrument exhibited slight color and spherical residuals. Later, OSLOT-LT 5.4 became available and the author downloaded this powerful free software. After entering the system parameters, the problem was found to be the tiny (2-inch diameter) singlet corrector's inherent strong spherochromatism when pushed to *f*/10.6. It was further determined, by ray trace analysis, that aspherizing the reflecting side of the corrector was not a good approach as spherochromatism would be roughly halved if the front (refracting) surface were aspherized instead. The lesson here is to always aspherize the refracting surface(s) of singlet or multi-element Schupmann correctors when necessary. This instrument's construction features are further detailed in Chapter 6. In early 2002 the objective was removed and, along with a new tube and 4.5-inch diameter singlet corrector, make up the authors present medial. A photo of this new instrument is shown in Figure 12.12. The remaining split-field parts, including the tube, are currently stored.

Twelve year's experience with the split-field concept has encouraged the author to make provisions for both split-field and tilted component operation in his next medial, a 14.1-inch *f*/12.5 presently under construction.

Fig. 12.12 *Jim Daley's a 9-inch f/11.1 medial.*

12.12 Jim Daley's 9-inch *f*/11.1 Medial

Designed and built in 2002, this instrument is used almost entirely for CCD double star measurements. This medial employs a half-diameter (4.5-inch) corrector. The tube is made from inexpensive lauan plywood and pine with solid mahogany cell supports as outlined in this book. For most critical astrometric work the tube is loosely covered with aluminum foil or sometimes aluminized Mylar to keep the internal air temperature from falling below the nighttime ambient. The foil was removed here to show the structure. Backyard tests show that using an aluminum - painted sample tube appears to be equally effective, so painting the tube aluminum is the next step. The idea, of course, is to reflect the night sky. Professional astronomers have caught on and are employing similar thermal control measures, the WIYN telescope's "dome" being one good example.

The mount is a standard German equatorial with a 4-inch diameter polar axle and a 3.5-inch diameter axle for declination. There is no drive in declination, just an adjustable friction pad, the telescope being simply hand-pushed to center up in the narrow field CCD crosshair centering eyepiece. Polar alignment is accurate and 30-second exposures of faint close doubles is no problem. The bearings are plain sleeve bronze bushings. To greatly reduce friction in RA a simple, easily removable, wooden counterpoise arm unloads much of the telescope weight and can be seen in the photo. The lifting force is provided by weights (not shown) that hang just off the floor on the end of the rope at the lower end of the arm. A fulcrum point gives a factor of two forcing gain. The upward force acts through a small ball bearing pivot at the very north end of the polar axis. Due to atmospheric effects as well as possible mount flexure reasons, all double star work is done fairly close to the meridian, say, plus-minus 1 hour, thus there is no interference from the counterpoise system. The telescope is never flipped west of the pier, each double being measured in turn near meridian passage, the north-south band of the instrument's astrometric position angle calibration, the CCD is precisely keyed in rotation—but measuring doubles is easily the subject of another book!

The 3-inch finder is a bare minimum aperture for acquisition of most neglected doubles. It can be seen clearly on the cover drawing of the instrument. No "go-to" computer drive stuff here, just a good declination circle, a synchronous motor worm and worm wheel RA drive and a nice smooth locking clutch. Guide-8 star chart print-outs, matching the 2.5-degree finder field, are all that's needed from computer land to quickly get on the object. Various other details of this telescope can be found throughout this book.

Little is said about observatories in this book, however, it should be clear to all that if scientific CCD astronomy is to be performed with a medial, or any instrument for that matter, some sort of permanent installation is required. Additionally, a small, attached, electrically heated and thermally insulated computer room is vital to long-term success.

12.13 The 1-meter Swedish Solar Telescope

Figure 12.13 shows the 0.97-meter clear aperture *f*/21 fused silica singlet objective being uncovered for the day's observations. A thin aperture ring shades the lens edge and "O" ring vacuum seal, preventing uneven lens heating. The lens edge is fine ground and buff polished to eliminate stress concentrations inherent in a standard edged surface, important when under vacuum load. The objective is carried in a two-mirror alt-az turret arrangement closely resembling that of the Dunn Solar Telescope at Sacramento Peak in New Mexico. (See *Sky and Telescope*, December 1969.) The chief difference in the two turrets being that here an objective lens rather than a plane window seals the evacuated optical path. Two 1.4-meter Zerodur optical flats, each at 45 degrees to the axis, direct the converging beam vertically down the tower to focus on the 60mm field mirror. Fused silica was chosen for the objective (and corrector) for its immunity to strain birefringence caused by center-to-edge thermal gradients. Strain birefringence is one of the solar astronomer's enemies as it can profoundly upset delicate polarization studies.

With the optical path evacuated, atmospheric pressure uniformly loads the 82.4mm thick objective with tremendous force which deforms the lens, however, detailed analysis carried out by the Swedish design team show near perfect cancellation of the first surface deformations, by those of the second surface. Because a highly corrected image is required at the first focus for direct observations in monochromatic light, the objective's first surface is figured to eliminate its weak

Fig. 12.13 *The objective end of the Swedish solar telescope (see Figure 1.10 for optical layout). Here, University of Oslo students have just removed the lens cover using a special handling fixture. Photo courtesy of John W. Briggs.*

Fig. 12.14 *Overall view of the Swedish solar telescope's turret. Photo courtesy of John W. Briggs.*

spherical aberration. Corning Glass manufactured the blanks, and the optical work for both the objective and corrector was done by Opteon Oy in Finland. Optical testing of the objective was performed by autocollimation off one of the 1.4 meter turret flats.

This medial may represent the final step in ground-based refracting solar telescopes. Although, a 1.5-meter version is feasible, the tremendous cost may outweigh the half-again improvement in resolution. Large Earth-based solar reflectors are in the works with a new 1.6 -meter unobstructed instrument nearing completion for installation at Big Bear Solar Observatory in Big Bear City, California. Located on Big Bear Lake, the fine seeing combined with the new instrument's adaptive optics promise to bring solar resolution to the next level.

Chapter 13
Making a 13-inch Schupmann by Scott Milligan

13.1 Some Background

Most amateur astronomers throughout the world are familiar with Stellafane—the Pink House, the Porter Turret telescope atop Breezy Hill, and the annual convention of telescope makers hosted there by the Springfield Telescope Makers.

Until the 1980s this small patch of land that Russell W. Porter donated in the 1920s to the Springfield Telescope Makers was all that made up Stellafane. In fact, Breezy Hill was so small that parking and camping facilities for the annual convention were in a nearby field which was "loaned" each year by its owner. But in the mid-1980s, the club was advised that the land would eventually become a tree farm. So, in 1987 to insure to continuance of the convention, the nearby 40-acre Darryl Sinon farm was purchased. This new land was promptly christened Stellafane East. In 1998, Harty and Peg Beardsley donated 45 acres of land adjacent to this property expanding Stellafane East to 85 acres. Harty Beardsley was

Fig. 13.1 *Early morning on Breezy Hill during the annual Stellafane Convention, 2005.*

Fig. 13.2 *The McGregor Observatory at Stellafane East which houses the 13-inch f/10 Schupmann telescope. Photograph courtesy Berton C. Willard.*

a member of the Springfield Telescope Makers and President of the Club in the 1950s.

While the Pink House and the Porter Turret Telescope "anchored" Breezy Hill, Stellafane East was barren of telescopes. In what follows, Scott Milligan, a member of the Springfield Telescope Makers, recounts the story of the fabrication of the optical elements for the first telescope to be installed on this new land—the 13-inch *f*/10 Schupmann telescope which saw its first light in 1995.

13.2 An Introduction to the Schupmann Telescope at Stellafane

It was 1984 and my second Stellafane. As I strolled around I came across two gentlemen who were busily engaged in setting up a rather strange looking wooden-tubed telescope. It turned out they were preparing to view Venus in daylight. While waiting for my first-ever daytime glimpse of Venus the conversation turned to their unusual telescope. "It's a single-glass refractive apochromat," one said, "and it uses only spherical surfaces." If I would stick around I could see for myself that its residual color aberration was essentially zero!

For those not familiar with recent Stellafane history, the telescope was a six-inch Schupmann built by Mike Mattei of the Amateur Telescope Makers of Boston. It had won first prize for optical excellence at the 1983 Stellafane convention. His companion was Jim Daley, whose self-published booklet *Amateur Construction of Schupmann Medial Telescopes* had inspired and guided Mike's efforts.

Working in collaboration with fellow STM member Bert Willard, who is also an optical engineer, Jim developed and prototyped the design that would eventually inspire the efforts of the Springfield Telescope Makers to build the world's largest operational Schupmann telescope.

The seed that would eventually become McGregor Observatory was planted in June of 1987. That month, I decided it was time to strip and recoat the 10-inch mirror that had served as my gateway to the stars for two years. After receiving some advice that suggested the use of a rather strong aqueous solution of sodium hydroxide to accomplish this task, I set up such a solution and immersed "my precious." Within 10 minutes most of the old coating was gone; a few stubborn spots, however, remained untouched. I then made a decision which, in hindsight, proved to be both pivotal and stupid. I decided to leave the mirror immersed overnight in the stripping solution. The next morning, I found the mirror to be completely free from aluminum deposits; unfortunately the polished optical surface had also been quite thoroughly etched, and was now totally useless as an objective!

Working at a determined pace to refigure the mirror in time for Stellafane, my thoughts drifted towards dreams of owning a large refractor. At that moment, the primary virtue of such an instrument was seen to be the absence of finicky mirror coatings; issues like central obstruction, tube currents, and coma free fields of view were minor considerations when compared with the benefit that I would never have to strip and recoat anymore #$%&$#@ mirrors! A problem with large refractors, however, is the presence of secondary spectrum. Although I knew observers who claimed not to be bothered by residual color, I personally found the presence of the bright violet halo to be an objectionable distraction. I considered the various options, including the use of special dispersion glass, and the various all-reflective TCT (tilted component telescope) designs. Use of special glass was attractive from a theoretical standpoint only, since the exorbitant cost (coupled with the substantial risk of breakage in relatively inexperienced hands) made the construction of a large instrument a test of financial courage. All-reflective TCTs, while in vogue among advanced ATMs, had a reputation for being somewhat difficult to collimate, had an awkward appearance, (yes, appearance does count for something!), and most important, did not solve the "freedom from mirrors" problem. Only the Schupmann, with its unobstructed, primarily refractive, and yet color-free optical system offered the possibility of satisfying most of my wish list. A further advantage of the Schupmann concept was the incredibly detailed yet clearly explained instructions for building the design contained in Jim Daley's booklet. Given the depth and clarity of the information presented, I felt confident that I could successfully complete the construction of a large Schupmann telescope.

13.3 What is a Schupmann?

Figure 13.3 shows an optical layout of the Daley-Willard-Ashcraft all-spherical Super-Schupmann. Light arriving from the heavens first encounters the main objective lens, labeled L_1 in the figure. The light then is focused to an intermediate

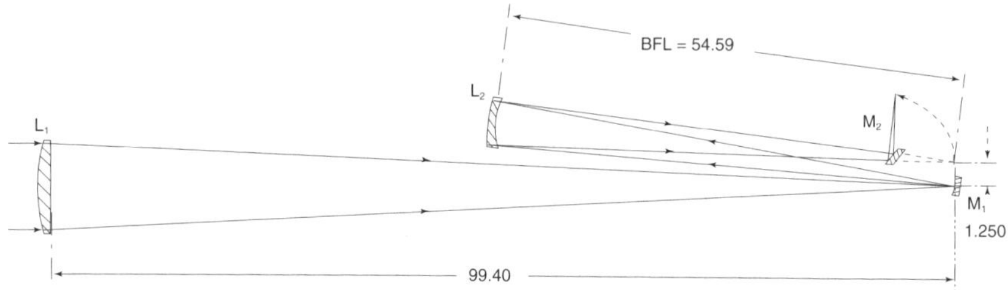

Fig. 13.3 *Optical Layout: Daley-Willard-Ashcraft Super Schupmann.*

image at the field mirror, labeled M_1. The intermediate image is aberrated, containing both under-corrected spherical and chromatic aberrations.

The field mirror M_1 has a weak concave curve, and reflects the incident radiation out to the corrector lens, L_2. The corrector lens is a negative meniscus refractive component, which has a reflective coating applied to its rear (convex) surface. The negative element is made from the same (BK7) glass as the main objective, preferably from the same melt, and the power and aperture of the negative element are chosen to correct the chromatic and spherical aberration introduced at L_1.

The reflective coating creates a concave mirror, which serves to return the light back through the negative meniscus and on its way to a small fold mirror, M_2, where the corrected image is directed towards an accessible focal plane. The field mirror M_1 by virtue of its weak positive power serves to re-image the aperture stop at L_1 onto the corrector L_2. This allows for complete correction of lateral color, and improves the already superb correction of the axial chromatic aberrations by equalizing the ray height at which the various wavelengths arrive at the L_2 corrector.

A "Lyot stop" is machined into the corrector lens cell; it is this aperture which determines the actual aperture of the telescope. Together with an adjustable field stop at M_1, the Lyot stop creates the potential for a fully-baffled optical train. The only possible unwanted rays are "straight shots" emanating directly from L_1 to M_2; these are blocked by baffles inside the tube.

To allow access to the final corrected image, this image must be laterally separated from the uncorrected intermediate image at M_1. This is accomplished by introducing a slight tilt in the plane of Figure 13.3 at the L_2 corrector. Introducing this tilt has the unfortunate consequence of introducing astigmatism into the final image. This astigmatism may be removed by several means; by far the simplest and therefore most likely to succeed technique is to introduce an offsetting tilt at L_1. This tilt must be orthogonal to the tilt introduced at L_2 if it is to correct the astigmatism introduced by the tilt in L_2.

A friendly word of advice for anyone wishing to set up a ray trace of the design: whatever tilt you select for the field mirror, you will probably find, as I did, that it is necessary to add a minor "tweak" to the angle that defines the location of the L_2 corrector after reflection at the field mirror. I found that optimum performance was achieved by tweaking both the x and y tilt angles here. Without the tweak, some lateral color remains on-axis. Remember also that the tilt of the objective lens must be introduced in a plane that is orthogonal to the L_2 tilt, or else the objective tilt will add its own astigmatism to that already introduced by the tilt of L_2.

13.4 Assembling the Team

The first hint that I had perhaps crossed over the invisible line separating dream from reality was during July of 1987, when I first shared my enthusiasm for the Schupmann concept with co-worker and fellow ATM Phil Rounseville. Phil and I had first met in 1983, shortly after I had joined the Amateur Telescope Makers of Boston. We had since been working together at Optical Systems and Technology, in the demanding and sometimes frustrating business of optical fabrication and testing. We quickly discovered our mutual interest in telescope making, and astronomy would often keep us at the shop after regular working hours had passed.

At first, I envisioned the project to be an individual effort. Almost immediately, Phil expressed his enthusiasm, despite the fact that neither one of us had the slightest idea as to where such an instrument could be permanently housed, or what it would cost. Ignoring these practical difficulties for the moment, we outlined the course of our dream.

The first imaginary dragon slain was the question of aperture. We both agreed that ten inches was the absolute minimum to consider, since anything less would probably not give enough image brightness at the really high magnifications that the Schupmann design was purported to excel at presenting. But was ten inches really the optimum aperture? Jim Daley's design worked at *f*/10: the largest optical test table available at Optical Systems and Technology was twelve feet long. Clearly, a 12-inch *f*/10 instrument would fit on this bench. Having read that the world's largest currently operational Schupmann was a 12.75" instrument housed in Stuttgart, Germany, spurred us to consider the possibility of building ours just a silly millimeter larger. In our minds, we had already made the leap from 10 to 12 inches; the fact that the possibility of owning the world's largest Schupmann presented itself as a mere sidestep from 12 to 13 inches meant that the die was cast: were we to actually build this telescope, it would be 13 inches in aperture.

Harry Vandermeer, President and founder of Optical Systems and Technology, offered immediate and generous support for our plans. Not only did he arrange to allow us use of OSTI facilities after working hours, but he also secured from Schott Glass Technologies, Inc., a donation of the raw glass blanks for the optical elements. OSTI also supplied material and labor hours to make the necessary optical tooling we would require to do a professional job. The only condition placed on our acceptance of this largess was that the finished telescope would become the property of a non-profit organization. The Amateur Telescope Makers of Boston had already agreed to spend up to $3,000 to fund construction of the tube assembly, and the Springfield Telescope Makers had decided to construct an observatory in memoriam of a recently deceased member, Douglas McGregor. Phil and I decided to begin the optical fabrication work as soon as raw materials could obtained.

13.5 Fabricating the Optics

After contracting with a local machine shop to make aluminum holders for the lens elements and pitch laps, we began the tedious task of making the test plates to check the radius and figure of the convex surfaces. Although Daley's booklet points out that only one concave test plate is needed to build the design, (for the first surface on the main objective), we theorized that having test plates for all of the convex surfaces would allow us to pinpoint the exact nature and location of residual figure errors without ambiguity. In order to test as large an area of each aperture as possible, we decided to make the test plates for the main objective out of 10-inch full-thickness Pyrex blanks, while the test plate set for the corrector convex would be 8-inch diameter Pyrex.

Phil and I began by fine grinding the convex and concave curve—generated test plate blanks against each other. As fine grinding progressed, frequent use was made of spherometers to ensure that the radius of curvature was accurately maintained. We found that with some practice, spherometer readings were repeatable to within 0.0001–0.0002 inch.

Care was exercised at this stage, lest the grinding slurry become too thin, which could cause the glass disks to cement themselves together. (The classic technique of grooving the grinding tool is not possible, since the "tool" is to become one half of the test plate set.) After progressing through a 5-micron fine grind with white aluminum oxide lapping powders, pitch laps were made on aluminum tools, and a quick polish was given to the concave member of each test plate set. Once polished, the radius of curvature was accurately measured, using either an instrument called a "radius bench," or a Ronchi tester and tubular inside micrometer. Subsequent adjustments to the radius of curvature were then made by varying the geometry and speed of the polishing strokes.

When the concave radius measured accurate to within plus or minus 0.005" (the long radius of 763 inches on the second side of the main objective was measured with a repeatability of plus or minus ⅜ inch), the concave plate was set aside, and the radius of the convex plate was adjusted to match the concave. Once the radii of the two matching plates were equalized, the surface figure was smoothed by working one half at a time, so as not to lose control over the radius. Figure 13.5 shows a 10-inch diameter test plate set under evaluation during this process. Quasi-monochromatic light is used to enhance the visibility of Newton's fringes.

During the in-process testing of the large diameter test plates we noticed that the amount of time for the glass to reach thermal equilibrium was always at least 30 minutes. Prior to reaching equilibrium, the appearance of the Newton fringes

Fig. 13.4 *Checking the 8-inch test plate with a spherometer.*

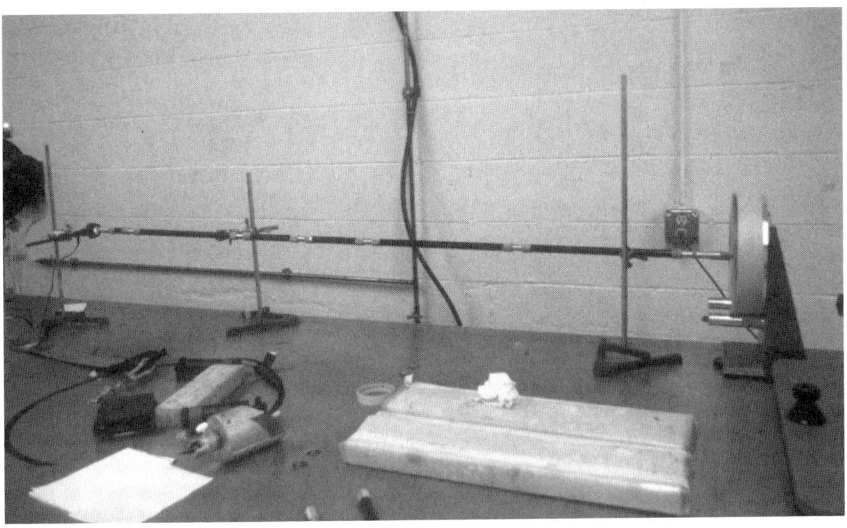

Fig. 13.5 *A radius bench. Precisely measured rods are assembled to match the desired radius of curvature.*

would vary in number and in shape. Most vexing were localized distortions which indicated either true errors in surface figure or the presence of thermal gradients across the glass disks. We also noticed that the weight of the upper disk was sufficient to introduce astigmatism into the match, unless care was taken to "float" the glass stack on foam rubber pads, in much the same way that a well-designed

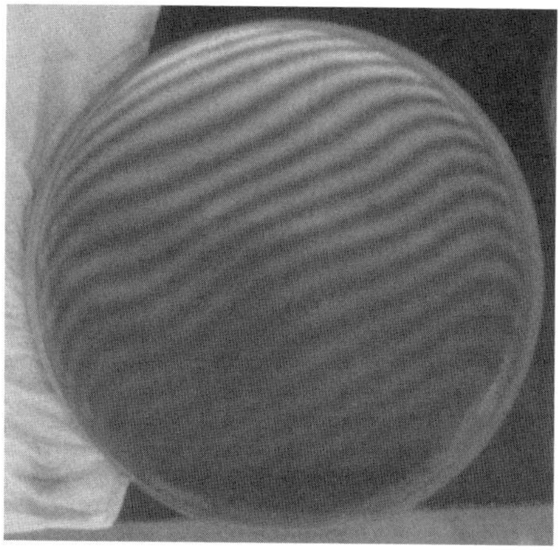

Fig. 13.6 *A 10-inch test plate set in the interferometer.*

Fig. 13.7 *Fine grinding the big lens.*

mirror cell "floats" its objective.

Altogether, the time spent making three sets of test plates probably amounted to about 300 man-hours. This alone is reason enough to heed Jim Daley's advice and make only one concave plate for the outer surface of the main objective. Additional reasons to follow this course would soon become apparent.

With the test plates out of the way, work commenced on fabricating the lens elements. Both the lens blanks and the Pyrex grinding tools had their curves

Fig. 13.8 *Measuring wedge on the corrector.*

roughed in on a curve generator, which saves much time and elbow grease. Fine grinding then began in the usual way. To attach the lens blanks to the rotating spindle, pads of ¼-inch thick foam rubber were inserted between the glass and the curved surface of the corresponding aluminum lens holder, and the lens was then secured to the holder by lightly wrapping the edge with 2-inch wide masking tape. Care was taken to insure that the tape was not wrapped so tightly that the glass blank was being pulled downwards at its edges. The generated Pyrex grinding tools for each surface were then grooved using a diamond cut-off saw, and the tool and lens edges were beveled before any grinding began. The diameters of the grinding tools for convex surfaces were sized about 90% of the lens surface diameters; the tools for concave surfaces were oversized by about the same amount. This arrangement allows work to progress with the concave surface (whether tool or lens) always on top, without giving up the ability to control radius of curvature.

As soon as the curve generating marks had been removed from one side, the lens blanks were measured for wedge error and center thickness. Wedge errors were marked for orientation, and then removed by applying hand pressure to the high side of the wedge with the lens on top and the spindle rotating at 10–15 rpm.

13.6 A Word Concerning Tolerances

At the time when tolerances needed to be set for our project, I had neither the knowledge nor the analysis tools to investigate the sensitivity of the Schupmann design; I therefore played it safe, and established tight tolerances on every construction parameter, to assure a successful outcome. Having since gained some ex-

Fig. 13.9 *Phil with a freshly made lap for L_1.*

perience in the field of lens design, I decided to perform a hindsight-aided tolerance sensitivity analysis.

Allowing for the use of suitable compensators, I found that radius of curvature errors of plus or minus 0.1% on all surfaces produce essentially no image degradation; these tolerances could, in fact be doubled without harmful effect. Wedge errors of 0.005" are easily tolerated (because the design is optically "centered" at the adjustable field mirror), and the center thicknesses can be off by 0.05" to 0.10" (or even more in the case of the big lens). Try building a conventional doublet or triplet to these tolerances, and you will quickly find that image quality drops to zero. In summary, I found that the only tolerances which really matter to the Schupmann are the figure accuracies and smoothness of the optical surfaces! This is particularly true of the reflective (convex) surface of the L_2 corrector, where a given figure defect, because of the in-glass reflection, introduces a corresponding wavefront error of almost six times greater magnitude than would be the case if the same defect were located on one of the refracting surfaces.

13.6.1 Polishing

Once the fine grinding operation had been completed for both surfaces of a lens, pitch laps were poured onto the same radiused aluminum tools that had doubled as lens holders during the fine grinding operation. A mixture of roughly 2 parts "Gugolz # 64" to 1 part "Gugolz # 73" pitches was used; the working temperature was always between 68 and 72 degrees Fahrenheit. The laps were poured onto the heated aluminum disks, and after a suitable cooling period, were pressed against the corresponding (wetted) glass surface in the usual way.

Fig. 13.10 *Scott holding the main objective lens.*

Since the definition of a "suitable" cooling period varies with many factors, I'm afraid that there can be no better guide than experience to judge when the lap is ready for the initial pressing. Groove patterns were laid out using a grease pencil and straightedge, and then the grooves were cut-in using a razor blade. No rubber mats or pitch "molding" techniques were used, however we did find that irregular fracturing of the groove edges could be minimized by cutting the laps under water. We also found that the quality of the contact made by the lap and (particularly) the large lens during polishing could be markedly improved if the pitch facets were either divided in two (by cutting additional grooves), or else by *lightly* scoring the pitch facets diagonally with a razor blade. Pitch thickness of the finished laps was about ¼" for the big lens, and maybe a bit less for the corrector.

13.6.2 Polishing Continued

Polishing continued until all signs of pits were removed. Now it was time to measure radius of curvature. The concave surface of the corrector posed no unusual problems, and was accurately measured on the radius bench. The convex surfaces were then compared against the concave halves of the recently finished test plate sets.

Here, I must digress for a moment to recall that there is nothing quite like the feeling one gets in the pit of the stomach while lowering a 13-pound full-thickness Pyrex disk onto the naked and defenseless surface of a large and (in your eyes) priceless optic. The only other sensation that remotely matches this one is the feeling you get while engaged in removing said test plate after the test is complete! It would be hard to overestimate the amount of force that must be applied, in a perfectly controlled manner, to achieve this second, unwanted task. If all that I have said to this point is not enough to turn the reader against the use of mammoth test plates, consider this: after all is said and done, the amount of additional, reliable information that is gained by using a test plate larger than 6 or possibly 8 inches in diameter is (almost without exception) exactly nil. That is because the increased weight and thermal mass of the larger plate introduce distortions in the test patterns that more than offset the largely theoretical advantage of greater aperture coverage. These distortions were large enough that examination of the fringe patterns during the fabrication process could not reveal whether the residual errors were due to test-induced distortions, or were actually in the glass. Further progress could only be made by conducting an optical test of the entire system.

13.7 Mechanical Parts for the Tube Assembly

Almost as soon as Phil and I had decided that the project would go forward, I began laying out the tube assembly, and producing detail drawings of the mechanical parts. One of the first tasks was to design lens cells to hold the optical elements.

The basic layout for the field-mirror and corrector cells closely follows the schematics shown in Daley's booklet, with three fine-adjustment screws opposed by tension springs at the corrector cell, and two orthogonal adjustment screws and a center pivot at the field mirror. 6061 T6 aluminum was chosen as the material for its ready availability, ease of working, reasonable cost, and good strength-to-weight characteristics.

The inside diameters of the cells were oversized to allow for insertion of plastic shims, which would prevent metal-to-glass contact. The faces of the internal shoulders and of the retaining rings were to be eventually covered with strips of self-adhesive Teflon tape for the same reason. The rings were designed to attach with screws, to avoid the problems that can occur with trying to match large diameter threaded rings and cells. Stainless steel inserts were added to prevent the relatively hard ends of the steel screws from galling the aluminum cells.

Through word of mouth I had learned that the Amateur Telescope Makers of Boston possessed an unusually resourceful member and long-time telescope nut by the name of Hal Robinson. When I called to ask whether he would be interested in helping our little project, his answer was "It sounds interesting. Why don't you bring the drawings down to my house and we'll look them over together?" After seeing the drawings, Hal wasted no time in making suggested improvements, and began planning how he would use the machinery in his basement shop to make the parts.

The cells for the corrector and field mirror were straightforward jobs for his milling machine, but the cell for the large lens was bigger than his existing lathe could handle. "But you know, I've been working on rebuilding another lathe," he said. "It's over a hundred years old, but I have a 24" diameter face plate for it, and

Fig. 13.11 *The L_1 cell on Hal's lathe.*

if I just jack the head stock up about six inches, I could swing up to 24" diameters." Figure 13.11 shows the results of Hal's handiwork. In order to produce accurate reference surfaces that could be clamped to the face plate, the part was first rough machined, and then the external surfaces front and rear were lapped flat on a large flat iron grinding tool. The part was then returned to the lathe, and the final dimensions were accurately cut. By spring of 1988, the lens cells were finished.

13.8 The First System Test

An autocollimation test of such a large optical system posed the practical difficulty of finding a vibration-isolated test bench large enough to host the test. Fortunately, such a bench existed at OSTI, and with proper planning, we figured that our setup could be built, evaluated, and torn down in a matter of a day, or at most, two. The bench was an old cast iron lathe bed that had a 12' long x 4' wide 1" thick aluminum plate fastened on top of the ways. This arrangement was far from ideal from a vibration isolation standpoint, but it was the only available test space large enough to do the job. I quickly constructed makeshift plywood test stands to hold the optical elements that were mounted in their cells. The convex surface of the L_2 corrector was given a temporary silver coating, and the lens elements were loaded into their cells.

When the alignment of the first system test had been optimized, a rather severe three-point astigmatic pattern was observed in the interferogram. At that juncture, we did not have any conclusive evidence which pinpointed the surface or surfaces that were to blame, but it was logical to suspect the convex surface on the L_2 corrector, since this is by far the most sensitive surface in the design. To check our assumption, the L_2 lens was rotated 60 degrees, and when the three-point pattern was also observed to rotate, we concluded that the source of the problem had been identified. But how could we fix it?

We were unable to leave the system test standing on the bench, and the test plate match was not sensitive enough to judge when the convex surface of the corrector was good enough. We needed another test that would be simple to implement, would occupy little space, and would give unambiguous results. Enter the Ross null test. Peter Ceravolo and Dick Stoltzman had just recently published their thorough investigation of this test for amateur use. Since the Schupmann's L_2 corrector lens, when tested by itself, created over-corrected spherical aberration similar to that imparted by a paraboloid tested at center of curvature, I reasoned that a Ross null could be found for testing the corrector lens. A few hours spent playing with the concept using Don Gibson's OSDP program confirmed my initial hypothesis; a two-inch diameter plano-convex lens made from BK7 would make a excellent null for our corrector.

Phil and I proceeded to make the null lens from a blank donated by OSTI. Meanwhile, Phil also continued to work on smoothing the figure of the convex surface of the corrector. He found that the extreme figure accuracy required, combined with the relatively thin aspect ratio (approximately 11:1) meant that the lens could no longer be mounted to its holder using masking tape around the perimeter. Instead, the concave surface of the lens was mounted to the convex tool by forming a multitude of pitch "buttons", each about ⅜" diameter, which were arrayed across the face of the holder, and then the lens was pressed gently onto the buttons while they were still warm enough to conform to the surface and adhere to the glass.

At last the null lens was finished, and we prepared to have our first look at the corrector as a stand-alone item. After scrounging a few necessary parts, we found the Ross null test to be simple to set up, align, and interpret. Figure 13.12 (left) shows an interferogram taken without the null lens in place; the L_2 lens shows about 5 waves of spherical aberration. Figure 13.12 (right) shows the interferograms taken through the first null test; notice that a central defect is revealed, but that the three-point astigmatism has been removed.

13.9 Finishing the Lens

Convinced that we were on the right track, Phil and I focused all attention on finishing the corrector lens. After a few more polishing spells, we obtained the interferogram shown in Figure 13.13; the small residual astigmatism present was judged to be unimportant, since the final adjustment of the objective lens tilt during system assembly would be an effective compensator for that aberration.

Considering the small time investment needed to make the Ross null test,

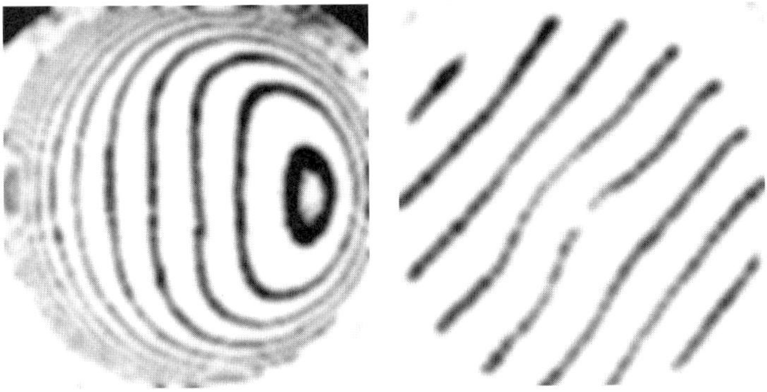

Fig. 13.12 *Interferograms without null lens (left) and with nul lens (right) showing central defect.*

Fig. 13.13 *Finished corrector in null test.*

and the high degree of figure correction it enabled us to obtain on a relatively difficult component, we concluded that Ceravolo and Stoltzman deserve some credit for clearing a path for our success in this area.

13.9.1 Hiatus

Meanwhile, construction had started on the building that would eventually become McGregor Observatory. I might have more to say on this subject, were it my story to tell, because as it begins, I was about to fall in love with the woman who would eventually become my wife. In fact, during the next three years I would become engaged, get married, change jobs three times, and build a house. Suffice it to say that my involvement with the construction of the observatory was minimal, and little or no forward motion occurred to the Schupmann optics during this time. But I became a happier person.

13.9.2 Resurrection

As the winter of 1994 approached, people began asking me "When will the optics be ready?" Since honor, and a desire to go on living, demanded that the answer to this question be "soon," I decided that it was time to tackle several tasks that were holding up completion of the optics. First among these was the need to build a white-light test device. The need for such a test device stems from the fact that both spherical aberration and chromatic aberration vary as the spacing between the L_1 and L_2 lenses is varied; this means that to get accurate correction for spherical aberration in the system test, the spacing must first be adjusted to eliminate color. I designed my tester to use a fiber bundle input, which is re-imaged onto a small hole that accepts pinhole or slit apertures.

The re-imaging optics are two collimating lenses; the collimated airspace between the lenses serves as a perfect place to insert various narrow-band filters, allowing the test device to function in either polychromatic (white light) or monochromatic mode. A right-angle prism scrounged at a Stellafane swap table is cemented to the plano surface of the second collimating lens; this places the body of the tester off to one side, and allows the observer's eye to get close to the image under test.

A second, vital piece of test equipment, which my frequent job changes had made no longer available to me was a vibration-isolated test bench. Repeated spasms of equipment envy had left me able to recite the brief tutorials on vibration-isolation systems in the Newport and Melles Griot catalogs almost by heart; from these, I learned that an imperfect, yet probably adequate isolation system could be constructed using nothing fancier (or more expensive) than a set of inner tubes and a pile of wood. Since the alternative was to take a second mortgage on my house and actually *buy* one of the systems that these companies are trying to sell via these tutorials, I decided that I had absolutely nothing to lose by taking the low road.

As it turned out, the table that I was able to construct cost of less than $300 and one weekend's labor, performed *better* for this task than some $20,000 setups that I had used or seen others use in the industry. Here's what I did: two 2 x 6s, 16 feet long form the side rails, which are separated by six 2 x 6s 4 feet long to create a frame 4 feet wide x 16 feet long. Sheets of ¾" thick AC plywood are attached to both sides of this frame via drywall screws (I wanted to be able to disassemble this beast later on). This structure forms the table, which floats on five 13" (inside) diameter inner tubes. The whole thing is supported by five 50 gallon drums, into each of which 100 lbs of sand has been poured to dampen vibrations. Plywood disks cover the drum tops. Figure 13.14 shows the result.

Using this set-up, we were able to perform a second system test, this time knowing that the L_1 to L_2 distance was correctly set for zero color. A microscope yielding a magnification of 100x was used to view the *f*/10 Airy disk in autocollimation. With the spacing adjusted for zero color, a LUPI (graciously lent by John Buccini of Buccini Instruments) was used to evaluate the double-pass interferogram. This test showed that although overall smoothness and symmetry were good, there was about ¼ wave of something that could best be described as overcorrected spherical aberration still present.

Fig. 13.14 *Homemade test bench—it works!*

Phil and I resolved to fix this last problem by figuring the front surface of the main objective lens, using a full-size lap and the "W" stroke familiar to all who have made parabolic mirrors by this method. New laps were made, and figuring was done in short spells of 5–10 minutes' work, followed by a look at the system test. After about 3 or 4 such spells, the "seagull" figure error was gone. Replacing the LUPI with the white-light source and the microscope, we saw that the double-pass Airy disk showed ring structure at the in-focus position, and the out-of-focus Airy patterns exhibited good symmetry on both sides of focus. The only noticeable image defects were a slight astigmatism, which we were confident would be taken care of by adjustments to the final alignment in the tube assembly, and a brightening of the outermost region in the Airy pattern inside of focus, indicating a slight up-edge in the wavefront. We decided to stop work, and begin integrating the optics into the tube assembly.

13.10 Stuffing the Tube

In October 1994 we began to stuff the tube. Prior to then, I and other STM members were busy designing and making the small mechanical parts for the rear end of the tube assembly.

Jeff Lowe, Chris Houghton, and Arthur Crowe had designed a mechanism that would allow the observer to switch between two different focusers, without losing collimation. The design had flowed down from my wish list and overall layout, through Chris's assembly and detail drawings, and into the machine shops of Jeff, Art, and Dave Kelly, who had made the many small parts that would make the mechanism work. Meanwhile, I prepared the existing aluminum parts for anodizing by sanding the unmachined surfaces with an orbital sander. That Friday

Fig. 13.15 *Figuring the big lens on a homemade polishing machine.*

evening, Tom Spirock, Jim Flood and Maryann Arrien arrived from New Jersey, with the monster wooden tube loaded into the back of Tom's truck.

As Saturday dawned, spirits were running high, which was fortunate, since our very first task would prove the need for teamwork and enthusiasm. Assembling the metal parts that formed the rear end of the tube was a necessary first step, since the optics could not be tested in the tube without providing a means to hold the small folding mirror which allows unobstructed access to the final image. It wasn't long into this effort that we discovered that some of the parts didn't fit together "quite right."

A quick review of the tools available in my basement revealed nothing suitable for metalworking, and so Maryann volunteered to drive the errant parts an hour west, where Dave Kelly had the facilities to fix the problem. Meanwhile, the rest of us proceeded with assembling the parts which did fit, cleaning and loading the optics into their cells, and attaching the mounted optics onto the tube assembly. A few hours later, Dave and Maryann returned, and it wasn't long before we discovered other metal parts that needed modification. Dave once again stepped in to save the day, machining the necessary changes into the aluminum using a carbide blade on my table saw!

After the focuser cage was assembled, we proceeded to install the bulkhead

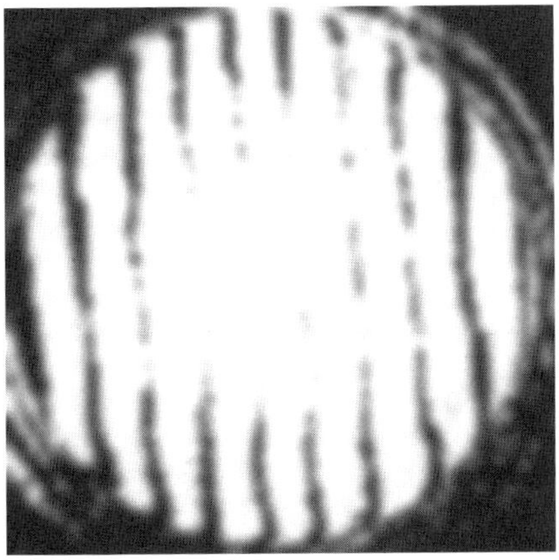

Fig. 13.16 *Double pass interferometer system test results.*

that would mount the main objective lens. To get the spacing between the objective and the M_1 field mirror right, we cut a wooden dowel to the length that would span the distance between the field-mirror cell and the inside corners of the tube where the bulkhead would be mounted. One end of this rod was held against the center of the field-mirror cell while the other was swung from corner to corner at the front end of the tube, so that the four corners where the bulkhead would attach could be marked. This technique proved remarkably accurate; the final alignment of the optics would require insertion of shims only 0.030 inch thick to set the tilt of the objective that would fully cancel on-axis astigmatism.

By late afternoon, the din and the dust began to settle, and we realized we were ready to begin aligning the optics. Bert Willard and Jim Daley had arrived to offer their assistance. A visual line of sight was used to collimate the optics, and the autocollimating flat was adjusted to return an image through the system. The white source and microscope were used as before to set the proper L_2 spacing for zero color. Finally, the LUPI was introduced, and the tilt of the M_1 mirror was adjusted to remove all traces of coma. Anxious faces hovered around the video monitor, waiting for the fringe pattern to settle down to tell us whether our efforts had been successful.

A collective shout of victory escaped ten mouths simultaneously when we realized that the fringes were straight and smooth across the entire aperture. As we studied the pattern in detail, however, we noticed something that looked like astigmatism. "OK," I immediately thought, "we need to adjust the tilt of the main objective." As we proceeded to do this, we noticed that our adjustments were having little or no apparent effect on the fringe pattern. Repeated attempts to remove the astigmatism by adjusting the objective tilt failed to improve the situation. Since it was getting late, it was decided to adjourn in favor of continuing the following day.

During the overnight hours, several of us realized that stratification of the air

Fig. 13.17 *The finished lenses in their cells.*

path inside the tube could introduce optical path differences which mimicked astigmatism, and that no amount of adjustments to the optical alignment would completely eliminate this problem. We began work on Sunday by opening the access doors that we had built into the tube. Using a portable fan, we began blowing air into the tube, hoping to create a thorough mixing of the stratified air layers that we suspected were causing the problem. Immediately upon doing this, the astigmatism in the fringe patterns was observed to diminish markedly. After Jim Daley pointed out that the effects of air path stratification would be less severe when the tube was pointed upwards (i.e. at the sky,) we decided to suspend further efforts at refining the alignment in the laboratory; instead, we would complete our alignment on the sky.

13.10.1 First Light

December 1994; the forecast was for overcast skies during the day, giving way to partly cloudy conditions overnight. I was relieved to notice that eight years of dedicated effort had not noticeably disturbed the overall climate for astronomy in New England. People would still want to live here because, unlike most places considered top-notch for astronomy, rain falls often enough that cactus and tumbleweed are successfully out-competed by real plants requiring non-fictional amounts of water to survive. Faith being everlasting, however, we prepared during the day for our first-ever views of the heavens through our new telescope.

The telescope tube and optics were disassembled for transportation to McGregor Observatory, which by now had a functioning roll-off roof and the rudiments of a computer controlled alt-az mounting that would carry the 270-pound tube assembly (and probably a great deal more weight than that) without batting an eyelash. The tube was hoisted ten feet above the floor using a chain-lift slung from the peak of the moving roof; the mounted optics were then carefully re-in-

Fig. 13.18 *Medial enthusiasts pose for a group photo alongside Stellafane's 13-inch f/10 Schupmann objective lens during the 2005 ATM convention. Clockwise from the left: Jim Daley, Scott Milligan, Clifton Ashcraft and Bert Willard. Jim, Scott and Bert are members of the Springfield Telescope Makers, Springfield, Vermont and Clif is very active in his club, the Amateur Astronomers Inc., Cranford, New Jersey. An overall view of this instrument can be seen in Figure 12.1. In this photo note how the hinged lens cover serves also as a sub-aperture solar filter holder. Photo by Perry Remaklus.*

stalled. Collimation of the optical train was easily accomplished using the adjustment screws on the M_1, L_2 and fold-mirror mechanisms. Our first terrestrial views confirmed the ability to reach focus with various eyepieces, and provided the first chance to experience tuning the lateral color in and out using the M_1 tilt screws. Now, all that was needed was some cooperation from Mother Nature during the overnight hours.

That night, we did get a few chances to view the sky. The telescope tracking and object acquisition software was not yet fully functional, and so we decided to do our star testing on Polaris, which then proceeded to dance in and out of the clouds all night. After tweaking the L_2 spacing to remove a hint of axial color, we observed that the images of Polaris were completely free from detectable color and showed no trace of spherical aberration. The turned-up edge alluded to earlier was clearly visible however, as well as a very slight astigmatism.

Problems with clouds and control of the telescope prevented going after either of these problems that evening, but we were satisfied that the basic level of optical quality was quite high. Before calling it a night, we turned our gaze to Saturn. Despite an unfavorably low position near the horizon, images of Saturn at 150x were crisp and quite contrasty, with little of the annoying glare that can accompany the light diffracted by an obstructed aperture. We agreed that we could finally afford to turn our attention away from the optics during the upcoming winter, and concentrate instead on readying the still unfinished Observatory building for final inspection and dedication in the early summer of 1995.

13.11 Current Status, and Plans for The Future

At the present time, the Observatory has been dedicated, and the telescope control system improved to the point where two experienced users can acquire and track objects with only a modest amount of "fudging" the system. Residual astigmatism at the center of the field has been removed by shimming the objective lens tilt, and the turned-up edge has been hidden by masking off the outer ¼-inch of aperture. Review of the test data suggests that it may be possible, with an accurate mask, to work with all but the outer ⅛ inch of aperture.

Currently planned tasks for the immediate future include installation of a precision aperture mask, internal light baffles, focuser upgrade and dew shield to the tube assembly. A parallel effort will address improvements to the telescope control and guidance systems, with near effortless operation by a single observer the eventual goal. Once these improvements are in place, the club plans to acquire a cooled CCD camera, and begin an imaging program. Images acquired with the telescope will be posted on the Stellafane web site, and on the Stellafane home-page.

Scott Milliagan gratefully acknowledges the selfless support for this project given by Mr. Harry Vandermeer, president, and the employees of Optical Systems and Technology, Inc. He also thanks Mr. Joseph Nimmerfroh and the Schott Glass Company for their generous donation of the glass for the Schupmann lens elements, and the Amateur Telescope Makers of Boston, who agreed early on to fund construction of the tube assembly. Finally, the author congratulates the membership of the Springfield Telescope Makers—many of whom labored for years on the construction of the McGregor Observatory—for their spirit of dedication, perseverance and commitment, without which this project would have proved impossible to realize.

Bibliography

Baker, James G., "The Catadioptric Refractor", *Astronomical Journal*, Vol. 59, No. 2, March 1954, pp. 74-83.

Berry, Richard, "Of Stellafane Optics and the Art of Deep Sky Observing," *Deep Sky Monthly*, Winter 1983, pp.33–34.

Brown, E.B., "Gleanings for A.T.M.s" edited by Earle B. Brown, "Report on a Schupmann," *Sky and Telescope*, April, 1942, p. 20.

Buchroeder, Richard A., "Design Examples of Tilted-Component Telescopes," Optical Science Center University of Arizona, *Technical Report No. 68*, May 1971, pp. 4, 43–45.

Cox, Robert E., "Amateur Astronomers" conducted by Robert E. Cox, "Stellafane 1964 Highlights," *Sky and Telescope*, October 1964, pp. 204-205.

Daley, James A. Jr., *Amateur Construction of Schupmann Medial Telescopes*, privately published, 1984, 63 pages.

Daley, James A. Jr., "The Schupmann Club," *Sky and Telescope*, March 1983, p. 278.

Dunn, R. B. *High Resolution Solar Observations*, Sacramento Peak Observatory, AFCRL, Sunspot, New Mexico, USA, Space Research XII - Akademie-Verlag, Berlin, 1972.

Editors Report, American Astronomers Report. "Achromatic Correction for Refractors", *Sky and Telescope*, February 1951, p. 93.

Hartmann, W.P., Private communication, 1993.

Hellman, Carl A., "Gleanings for A.T.M.s" edited by Earle B. Brown, "The Schupmann Telescope" *The Sky*, September 1941, p.14-15.

Lyot, Bernard, "A Study of the Solar Corona and Prominences without Eclipses," *Monthly Notices of the Royal Astronomical Society*, Vol. 99, No. 8, June 1989, pp. 580-594.

McGregor, Doug, "Stellafane '83," Amateur Astronomers Conducted J.W. Briggs, *Sky and Telescope*, November 1983, pp. 444–447.

Mündler, M., "Über ein 12-zölliges Medial," *Astronomishe Nachrichten*, No. 5223, Vol. 218, April 1923, pp. 235–240.

Musson, W. B., *The Observatory*, Correspondence, Vol. 23, No. 296, September, 1900, pp. 350–352.

Olson, Edwin A., "A Schupmann for Amateurs," *The Maksutov Club Circular*, No. 119, September 1964, pp. 15–19.

Orrall, F.Q., Thesis, Harvard University, 1954.

Overbye, Dennis, "A Night on Breezy Hill," *Discover Magazine*, October 1983, pp. 84–87.

Riekher, Rolf, *Fernrohre und ihre Meister*, Verlag Technik, 2nd ed. of Riekher, Berlin,1990. pp. 230-236.

Rolf, Edwin, Abschlussbericht zum Forschungsauftrag, F2-119, "Der Verkürzter Refraktor," Rathenow, 15 May 1953.

Rush, Joseph H., "The Amateur Scientist Department," *Scientific American*, May 1958, pp. 130-138.

Rush, Joseph H. and Schnable, George H., "High Altitude Observatory's New Coronagraph and Spectrograph," *Applied Optics*, Vol. 3, No. 12, pp. 1347-1352.

Scharmer, Göran B., Bejelksjö, Klas, Korhonen, Tapio, Lindberg, Bo, Petterson, Bertil, "The 1-meter Swedish solar telescope," *Proceedings of SPIE,* Vol. 4853, 2003.

Schupmann, Ludwig, "Zeitschrift für Instrumentenkunde" XLI Jahrgang, pp. 212-219.

Schupmann, Ludwig, *Die Medial-Fernrohre, Eine neue konstruktion fur grosse astronomische instrumente*, B. G. Teubner, Leipzig. 1899.

Sky and Telescope, February 2003, page 16.

Sky and Telescope, November 2004 page 22.

Starken, C., "Variable Stars as Targets for Long-Term Coordinated Multisite Photometry", in *The Study of Variable Stars with Small Telescopes*, edited by John R. Percy, Cambridge University Press, 1986, pp. 165-179.

Wolter, Cristian and Merz, Rainer, "Gleanings For ATM's" conducted by Roger W. Sinnott, "The Neglected Schupmann Refractor", *Sky and Telescope*, March, 1983, pp. 273-278.

Index